Principles of
Catalyst
Development

FUNDAMENTAL AND APPLIED CATALYSIS

Series Editors: M. V. Twigg

Imperial Chemical Industries P.L.C.
Billingham, Cleveland, United Kingdom

M. S. Spencer

School of Chemistry and Applied Chemistry
University of Wales College of Cardiff
Cardiff, United Kingdom

PRINCIPLES OF CATALYST DEVELOPMENT
James T. Richardson

A Continuation Order Plan is available for this series. A continuation order will bring delivery of each new volume immediately upon publication. Volumes are billed only upon actual shipment. For further information please contact the publisher.

Principles of Catalyst Development

James T. Richardson
University of Houston
Houston, Texas

PLENUM PRESS• NEW YORK AND LONDON

CHEMISTRY

036356 5X

Library of Congress Cataloging in Publication Data

Richardson, James Thomas, 1928–
  Principles of catalyst development.

  (Fundamental and applied catalysis)
  Bibliography: p.
  Includes index.
  1. Catalysts. I. Title. II. Series.
  TP159.C3R47   1989   660'.2995   89-15947
  ISBN 0-306-43162-9

© 1989 Plenum Press, New York
A Division of Plenum Publishing Corporation
233 Spring Street, New York, N.Y. 10013

Printed in the United States of America

Alchemists

Powerful particulates precipitated in a pot
Colloidal clusters coalesced and aged
Hydrogels hardened and heated, filtered and formed
Notions nurtured for the needs of now

Less primitive perhaps than brothers passed
No need for magic, incantations, spells
But insight gathered from the past
And postulates proven by performance

While science smothers supplication
Humility and faith remain
Stones yield not gold nor life eternal
But catalysts for rich resources

—J. T. Richardson (October, 1982)

This book is dedicated to the memory of

Professor W. O. Milligan,

a brother passed, who taught me much about science,
supplication, and the way things are.

FOREWORD

Successful industrial heterogeneous catalysts fulfill several key requirements: in addition to high catalytic activity for the desired reaction, with high selectivity where appropriate, they also have an acceptable commercial life and are rugged enough for transportation and charging into plant reactors. Additional requirements include the need to come online smoothly in a short time and reproducible manufacturing procedures that involve convenient processes at acceptable cost. The development of heterogeneous catalysts that meet these (often mutually exclusive) demands is far from straightforward, and in addition much of the actual manufacturing technology is kept secret for commercial reasons—thus there is no modern text that deals with the whole of this important subject. *Principles of Catalyst Development*, which deals comprehensively with the design, development, and manufacture of practical heterogeneous catalysts, is therefore especially valuable in meeting the long-standing needs of both industrialists and academics.

As one who has worked extensively on a variety of catalyst development problems in both industry and academia, James T. Richardson is well placed to write an authoritative book covering both the theory and the practice of catalyst development. Much of the material contained in this book had its origin in a series of widely acclaimed lectures, attended mainly by industrial researchers, given over many years in the United States and Europe. All those in industry who work with catalysts, both beginners and those of considerable experience, should find this volume an essential guide. With its emphasis throughout on the scientific basis of catalyst development, it should be an excellent teaching text, especially for postgraduate courses in applied chemistry, chemical technology, and chemical engineering.

Martyn Twigg
Michael Spencer

Billingham and Cardiff

PREFACE TO THE SERIES

Catalysis is important academically and industrially. It plays an essential role in the manufacture of a wide range of products, from gasoline and plastics to fertilizers and herbicides, which would otherwise be unobtainable or prohibitively expensive. There are few chemical- or oil-based material items in modern society that do not depend in some way on a catalytic stage in their manufacture. Apart from manufacturing processes catalysis is finding other important and ever increasing uses; for example, successful applications of catalysis in the control of pollution and its use in environmental control are certain to increase in the future.

The commercial importance of catalysis and the diverse intellectual challenges of catalytic phenomena have stimulated study by a broad spectrum of scientists, including chemists, physicists, chemical engineers, and materials scientists. Increasing research activity over the years has brought deeper levels of understanding, and these have been associated with a continually growing amount of published material. As recently as sixty years ago, Rideal and Taylor could still treat the subject comprehensively in a single volume, but by the 1950s Emmett required six volumes, and now no conventional multivolume text can cover the whole of catalysis in any depth. In view of this situation, we felt there was a need for a collection of monographs, each one of which would deal at an advanced level with a selected topic so as to build a catalysis reference library. This is the aim of the present series, Fundamental and Applied Catalysis.

These books in the series will deal with particular techniques used in the study of catalysts and catalysis; these will cover *the scientific basis of the technique, details of its practical applications*, and *examples of its usefulness*. The volumes concerned with an industrial process or a class of catalysts will provide information on the *fundamental science of the topic, the use of the process or catalysts*, and *engineering aspects*. For example, the inaugural volume, *Principles of Catalyst Development,* looks at the science behind the manufacture of heterogeneous catalysts and provides practical information on their characterization and their industrial uses. Similarly, an upcoming volume on ammonia synthesis will extend from the surface science of single iron crystals to the design of reactors for the special duty of ammonia manufacture. It is hoped that this approach will give a series of books that will be of value to both academic and industrial workers.

The series will deal with both heterogeneous and homogeneous catalysis, and will include processes in the heavy chemicals and oil refining industries, the smaller-scale manufacture of pharmaceuticals, and various aspects of pollution control. The series editors would welcome any comments on the series and suggestions of topics for future volumes.

Martyn Twigg
Michael Spencer

Billingham and Cardiff

PREFACE

If this book sounds like "all you ever wanted to know about catalysis and were afraid to ask," then perhaps it is. It is written unashamedly for the newcomer to catalysis, the student, process engineer, research chemist, salesman, and manager. The philosophy of the book is based on a highly successful industrial short course, "Applications of Heterogeneous Catalysis," taught at the University of Houston for over 17 years. My experience with this course tells me that there is a constant flow of scientists and engineers entering the field who need this type of text.

My own experience of the past 35 years is testimony to the complexity of catalytic development: graduate degrees in physics under the direction of a colloidal chemist; 13 years of industrial research with Exxon Research and Engineering, learning the balance of fundamental and applied catalysis; and 18 as professor of chemical engineering, developing and using teaching techniques. I would hope that today's apprentices do not need such a varied path but find within existing disciplines the guidance and background with which to advance catalysis from an art to a science.

And yet it is no easy task to write a book for such a heterogeneous (no pun intended) audience. Physicists, chemists, and engineers seeking a foothold will find in these pages a little of all their crafts but, alas, not enough to satisfy the purists. The objective is to provide the necessary essence of each discipline, to educate, to instruct, and to supply enough coherence that, one hopes, those of us from different backgrounds can work together or at least converse with each other. For those wishing to advance into a deeper study of catalysis, this book will serve as a focal point and a launching pad, through the references quoted liberally in the text.

The specialist is not forgotten. I have learned that each of us has still much to discover. If this book sparks one creative thought in each of its readers, its purpose will have been realized.

Chapter 1 starts with a review of catalytic functions, a necessary prelude to the journey, to remind us what the catalyst is expected to do. Then follows a discussion of catalytic components. Impatient readers may enjoy jumping ahead at this point, but it will be necessary to return for more serious contemplation after the intervening chapters have been mastered.

Chapter 3 continues with the philosophy of catalyst development and seeks to place every step in perspective. The reader should seek his or her place in the scheme of things or perhaps identify tasks yet to be tried. This chapter is, frankly speaking, preaching, but with the purposes of imparting an appreciation for the role each of us plays.

Chapter 4, on catalytic materials, is a tough one. Attempting to cover the fundamentals of catalysis as they exist today in one chapter is perhaps presumptuous. But it must be done. I have tried to introduce the right balance of historical fact and modern thinking, knowing all too well that both will change. This is the chapter where many readers will feel cheated, but, I hope, also inspired to seek further knowledge. Chapter 5 continues with a suggestion of how to design a novel catalyst composition.

In Chapter 6 the subject gets serious—catalyst preparation. Enough detail is given to be instructive but at the same time to deliver the right degree of respect for the "art" and expectation in the "science." Many readers will find this discussion their first insight into the magic of catalysis.

Catalyst characterization as discussed in Chapter 7 is straightforward and contains no inspiration, just the necessary facts. Even those not personally concerned with conducting the tests should follow this material with interest, for all of us are concerned with measuring our achievements.

Chapter 8 on catalyst deactivation is lengthy, but this is where the catalyst designer displays his or her ultimate virtuosity. Including it was too good an opportunity to miss.

Appendixes provide a source of reference material that will prove useful to the newcomer progressing into the arena to flex his or her knowledge.

Above all, this book is a teaching vehicle and as such should be enjoyed. Do not devour it all at once, but take your time. Your comments will be seriously and graciously received.

I am grateful to colleagues over the years who have contributed to the material used and to all my teachers and students, from whom I have learned much. Special appreciation goes to the University of Houston for providing a sabbatical opportunity to write the book and to the University of New South Wales for the pleasant environment in which to do it.

James T. Richardson

Houston

CONTENTS

CHAPTER 1. CATALYTIC FUNCTIONS: WHAT THE
            CATALYST DOES

1.1. A Definition   1
1.2. Organization of Catalysis   6
1.3. Structure and Texture of Catalyst Particles   8
1.4. Steps in the Catalytic Reaction   11
1.5. Diagnosis of Catalytic Function   19

CHAPTER 2. STRUCTURE OF CATALYSTS: WHAT THEY
            CONTAIN

2.1. Catalyst Engineering   23
2.2. The Balanced Particle—A Compromise   23
2.3. Catalyst Components   26
2.4. Example—HDS Catalysts   37

CHAPTER 3. CATALYST DEVELOPMENT: HOW THEY
            EVOLVE

3.1. Catalyst Development   41
3.2. Process Need   41
3.3. Define Problem and Objectives   43
3.4. Catalyst Research   44
3.5. Catalyst Design   45
3.6. Catalyst Testing   46
3.7. Pilot Unit Testing   46
3.8. Process Design   48

CHAPTER 4. CATALYTIC MATERIALS: WHAT THEY ARE

4.1. Types of Catalytic Materials   49
4.2. Metals   50
4.3. Semiconductors   63
4.4. Insulators and Solid Acids   69
4.5. Concluding Remarks   82

CHAPTER 5. DESIGN OF CATALYSTS: HOW TO INVENT
            THEM

5.1. Methodology   83
5.2. Example—Oxidation of Methane to Formaldehyde   85
5.3. Computer Aids   92
5.4. Some Final Thoughts   92

CHAPTER 6. CATALYST PREPARATION: HOW THEY ARE
            MADE

6.1. Introduction   95
6.2. Single Active Components and Supports   95
6.3. Dual Oxides 106
6.4. Deposition of the Active Components 108
6.5. Extraction 120
6.6. Special Types 122
6.7. Commercial Manufacture of Catalysts 123
6.8. Catalyst Forming 127
6.9. Mechanical Strength 131
6.10. Legal Responsibilities 133
6.11. Some Final Thoughts 134

CHAPTER 7. CATALYST CHARACTERIZATION: HOW
            THEY ARE TESTED

7.1. Preliminary Remarks 135
7.2. Bulk Properties 136
7.3. Particle Properties 140
7.4. Surface Properties 157
7.5. Activity 171
7.6. Some Final Thoughts 183

CHAPTER 8. CATALYST DEACTIVATION: HOW THEY FAIL

8.1. Preliminary Remarks 185
8.2. Effects of Deactivation 185
8.3. Cause of Deactivation 191
8.4. Some Final Comments 221

APPENDIXES

Appendix 1. Reference Books and Journals on Catalysis 225
Appendix 2. Glossary of Common Terms in Refining and
             Petrochemicals 237
Appendix 3. Units and Nomenclature 243
Appendix 4. Catalytic Process Assessment 247
Appendix 5. Classification of Reactions 251
Appendix 6. Laboratory Recipes 259
Appendix 7. Catalyst Manufacturers 263

REFERENCES 267

INDEX 279

1

CATALYTIC FUNCTIONS
What the Catalyst Does

1.1. A DEFINITION

This chapter deals with what a catalyst does and what it can be expected to do or not do. We start the discussion with a definition:

> A catalyst is a *substance* that *increases the rate* at which a chemical reaction approaches *equilibrium* without itself becoming *permanently* involved in the reaction.

Key words in the definition are italicized. A catalyst is itself a chemical substance and as such becomes involved in the reaction, although not permanently. The chemical state of the catalyst is subject to all the rules of chemistry in its interaction with reactants but remains unchanged at the end of the reaction. Primarily, the catalyst accelerates the kinetics of the reaction toward thermodynamic completion by introducing a less difficult path for molecules to follow. Figure 1.1 illustrates this feature with an industrially important reaction, the synthesis of ammonia[1]:

$$N_2 + 3H_2 \leftrightarrows 2NH_3 \qquad (1.1)$$

Nitrogen and hydrogen molecules combine homogeneously (without a catalyst) at an extremely slow rate. Breaking their bonds to form reactive fragments requires large amounts of energy, leading to an activation energy of 57 kcal mole^{-1}. The probability of finding these fragments together is small. Spontaneous ammonia formation at moderate conditions is infinitesimal. The catalyst, however, assists dissociation through chemisorption and recombination with a series of surface interactions:

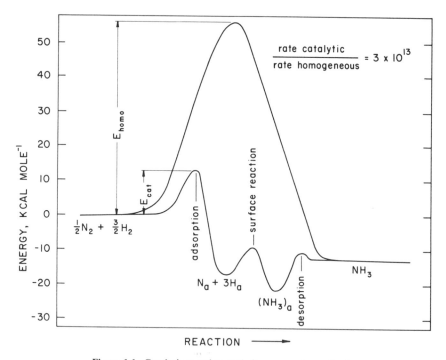

Figure 1.1. Catalytic reaction path for ammonia synthesis.

$$H_2 \rightarrow 2H_a \qquad (1.2a)$$

$$N_2 \rightarrow 2N_a \qquad (1.2b)$$

$$N_a + H_a \rightarrow NH_a \qquad (1.2c)$$

$$NH_a + H_a \rightarrow (NH_2)_a \qquad (1.2d)$$

$$(NH_2)_a + H_a \rightarrow (NH_3)_a \qquad (1.2e)$$

$$(NH_3)_a \rightarrow NH_3 \qquad (1.2f)$$

The rate-determining step is (1.2b), nitrogen adsorption, which requires an activation energy of only 12 kcal mole^{-1}. Rates accelerate enormously, by a factor of 10^{13} at 500°C. Notice that initial and final enthalpies are unchanged, so that equilibrium conversion, X_e, is the same. Conversions follow the relationships shown in Fig. 1.2.

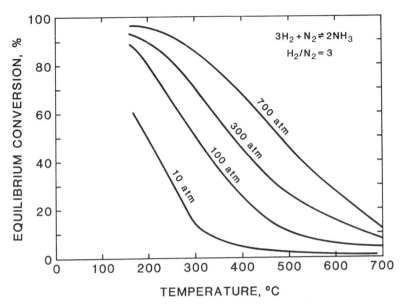

Figure 1.2. Conversions for ammonia synthesis.[1]

Our definition contains within it four implications that are the foundation for understanding the functions of the catalyst. First, as emphasized in Fig. 1.1, equilibrium is unchanged, together with all thermodynamic properties such as ΔG_r, ΔH_r, and K_r. The catalyst promotes only those reactions for which the Gibbs free energy change ΔG_r is less than zero. Feasibility is better defined by ΔG_r^0, which is not the same as ΔG_r, but is more easily calculated. Table 1.1 gives general criteria for whether or not a reaction is feasible.

TABLE 1.1. Criteria for Thermodynamic Feasibility

ΔG_r^0	Feasibility
1. Very negative $< -10\,\text{kcal mole}^{-1}$	Very high equilibrium conversions possible
2. Moderately negative 0 to $-10\,\text{kcal mole}^{-1}$	Fairly high equilibrium conversions possible
3. Moderately positive 0 to $10\,\text{kcal mole}^{-1}$	Low equilibrium conversions depending upon process conditions
4. Very positive $>10\,\text{kcal mole}^{-1}$	Very low equilibrium conversions, generally not significant reactions

Second, since the equilibrium constant, K_r, is unchanged, it follows from the equality

$$K_r = \vec{k}/\overleftarrow{k} \tag{1.3}$$

that the catalyst must accelerate both the forward rate constant, $\vec{k}$, and the reverse, $\overleftarrow{k}$. Although not a factor in irreversible reactions, this feature is important in appreciating the role of the catalyst in normally reversible situations. For example, materials that are known to function as hydrogenation catalysts will also be good for dehydrogenation, if compatible with the necessarily different process conditions.

Another more subtle point emerges. Sometimes studying a forward reaction is difficult, while the reverse is easy. An example is ammonia synthesis. This reaction is reversible over the range of temperatures normally encountered in industrial operations, 200–1200°C. Figure 1.2 shows that the exothermic synthesis reaction decreases in equilibrium conversion as the temperature increases. Higher yields are obtained by decreasing the temperature. But kinetic rates are lower, so precision suffers. Also, the stoichiometry of reaction (1.1) indicates that increasing pressure will raise the equilibrium conversion. In the case of NH_3 synthesis, this amounts to hundreds of atmospheres. High-pressure equipment is extremely inconvenient for most laboratory studies.

Ammonia decomposition, on the other hand, may be carried out under more favorable conditions. Stoichiometry favors low pressure, so normal atmospheric-pressure equipment is sufficient. Equilibrium yields increase with temperature and kinetic rates are measured with precision. This is why ammonia decomposition, which is less interesting, has historically received so much attention in the search for improved synthesis catalysts.[2]

The third implication from the definition is that more than one reaction may be involved, leading to different thermodynamically feasible products. A catalyst, in principle, promotes only one of these, leading to improvements in selectivity as well as activity. Since the catalyst is a chemical, reacting with reactants and products through chemisorption or complexing, its reactivity depends upon its own chemical structure. We see this demonstrated in the simple decomposition of formic acid:[3]

$$\text{Dehydration:} \qquad HCOOH \xrightarrow[Al_2O_3]{} H_2O + CO$$

$$\tag{1.4}$$

$$\text{Dehydrogenation:} \quad HCOOH \xrightarrow[Metals]{} H_2 + CO_2$$

More important industrial examples exist. For instance, by changing the catalyst (and process conditions) we may convert H_2 and CO mixtures selectively into methane (Ni/Al_2O_3),[4] paraffinic hydrocarbons (Fe/kiesel-guhr),[5] alcohols, aldehydes, and acids (Co/ThO_2),[6] or methanol (Cu/ZnO).[7] The catalyst becomes a useful tool for manipulating selectivity.

The last point concerns the question of permanent involvement by the catalyst. Ideally, the catalyst is unchanged by the reaction. In practice this is not true. Since it is itself a reacting substance, the catalyst suffers from irreversible chemical and physical changes, which decrease its ability to perform. Within the time frame of the reacting molecules, these changes are small. But as the process time continues and the catalyst experiences many billions of these events, deactivation becomes significant.[8]

We have extracted from our definition an appreciation for three catalytic functions: activity, selectivity, and deactivation. Which is the most important? This is difficult to answer generally since each application has its own set of specific needs.

Certainly, for the reaction to proceed, the catalyst must have chemical activity. Beyond that, increasing activity can have several benefits:

1. Higher rates for the same conditions.
2. Equivalent rates but with higher throughputs or smaller reactors.
3. Equivalent rates at lower temperatures or pressures where equilibrium yields increase, operations become easier, deactivation becomes less, or selectivity improves.

Selectivity becomes a factor in the presence of multiple reactions. These are generally of the types

Parallel:
$$R \xrightarrow{k_D} D \qquad\qquad (1.5)$$

$$R \xrightarrow{k_U} U$$

Or consecutive:

$$R \xrightarrow{k_D} D \xrightarrow{k_U} U \quad \text{and} \quad R \xrightarrow{k_U} U \xrightarrow{k_D} D \qquad (1.6)$$

With catalyst control, the ratio k_D/k_U may be increased to optimize the desirable product D. Benefits are obvious and include greater yields of D and less extensive separation operations. An especially important case occurs when U is a deactivating agent such as "coke" or carbonaceous deposits.

Although the importance of increasing selectivity is different for each process, in the case of industrial operations with large throughputs, small improvements may lead to large benefits. For example, a 1% increase in gasoline yield in catalytic cracking amounts to an extra one million gallons a day for the United States, thereby lowering dependence on imports.[9]

Deactivation rates enter significantly into process design. Not only are yield–time relationships established, but also type of reaction and mode of operation (e.g., fixed or fluidized bed).[10] Lifetime decline can have a profound effect on the engineering and economics of a process. Small changes that improve lifetime can have a large payoff. An example is the introduction of bimetallics in catalytic reforming. Adding rhenium to platinum on alumina greatly decreased the deactivation rate from both coking and sintering. Not only did the catalyst last longer and require less regeneration (thus vastly simplifying process configurations), but lower pressure operations were possible, giving improved selectivity to high-octane aromatics.[11]

Priorities in today's industry emphasize more efficient utilization of feedstocks and energy. Most effort is placed on improving existing processes rather than developing new ones. With this in mind and with an awareness of the direction of current research, it is the author's opinion that the relative order of importance is selectivity > deactivation > activity.

1.2. ORGANIZATION OF CATALYSIS

Before proceeding with our discussion, we should perhaps briefly acknowledge the different but complementary divisions of catalysis now existing within industrial and academic fields. They are homogeneous, heterogeneous, and enzyme catalysis. Each has its own structure and disciplines, involving researchers from varying backgrounds. There are specialized journals and meetings catering to each. Unfortunately, attempts at cross-fertilization have not been too successful, which is regrettable since each has much to learn from the others.

1.2.1. Homogeneous Catalysis

The catalyst is the same phase as the reactants and products.[12] Examples are hydrolysis of esters by acids (liquid–liquid), oxidation of SO_2 by NO_2 (vapor–vapor), and decomposition of potassium chlorate by MnO_2 (solid–solid). Usually, the liquid phase is most common, with both catalyst and reactants in solution.

Catalysis occurs through complexing and rearrangement between molecules and ligands of the catalyst. Reactions can be very specific, with high yields of desired products. Since the mechanisms involve readily identified species, these reactions are easily studied in the laboratory with the techniques of organometallic chemistry. They are, however, difficult to operate commercially. Liquid phase operation places restrictions on temperature and pressure, so equipment is complicated. The catalyst must be separated from the products, imposing additional difficulties.

For these reasons, homogeneous catalysis is found only in limited industrial use, appearing usually in the manufacture of specialty chemicals, drugs, and food. Exceptions are acetic acid production, olefin alkylation, and hydroformylation.

1.2.2. Heterogeneous Catalysis

In these systems the reactants and catalyst exist in different phases. Most commonly, solid catalysts are used with gaseous or liquid reactants, sometimes both. Other permutations are possible but less often encountered.

Introducing a separate catalyst phase immediately complicates the mechanism. Interfacial phenomena now become important. Diffusion, absorption, and adsorption all play critical roles in establishing the rate. These additional steps become difficult to separate from surface chemistry. Accordingly, heterogeneous systems are difficult to study in the laboratory. Disappearance of reactants and appearance of products are easily followed, but important features such as the rates and energetics of adsorption, structure of the active surface, and nature of reactive intermediates, require separate experimentation using a constantly changing arsenal of techniques. Often, the sequence of surface steps can only be deduced from accumulated evidence, with many uncertainties. In every important application of heterogeneous catalysis there is much controversy on the exact details of chemistry. Ammonia synthesis, for example, was the first important catalytic process, emerging over 70 years ago.[13] Yet the nature of the catalytic surface is still debated.[14] Another case is hydrodesulfurization. Process and catalyst development dates from the early 1950s[15] but a large amount of scientific research is now appearing in the current literature.[16]

Nevertheless heterogeneous catalysts are convenient to use commercially. Easily prepared solid catalyst pellets, packed in tubes through which reactants flow, satisfy process requirements for simple construction and dependable operation. Control is good, product quality high. It is not surprising that the vast majority of industrial catalytic processes adopt this approach.

1.2.3. Enzyme Catalysis

Enzymes are protein molecules of colloidal size, somewhere between the molecular homogeneous and the macroscopic heterogeneous catalyst. Thus they are neither but somewhere in between. Enzymes are, of course, the driving force for biochemical reactions. Present in life processes, they are characterized by tremendous efficiencies and selectivities. An enzyme, catalase, decomposes H_2O_2 10^9 times faster than any inorganic catalyst.[17]

There is great interest in harnessing enzyme catalysis for industrial use.[18] Much depends on increasing the ability to withstand severe conditions and developing techniques for anchoring enzymes to substrates for use in packed beds.

With these brief comments, we leave homogeneous and enzyme catalysis to others and concentrate the rest of this book on the development of heterogeneous catalysts.

1.3. STRUCTURE AND TEXTURE OF CATALYST PARTICLES

Manufactured as a powder, the catalyst is formulated into particles, whose shape and size are determined by the end use. Common types, with a brief description of each, are given in Table 1.2, with more details to follow in Chapter 6.

The first four types, pellets, extrudates, spheres, and granules, are primarily used in packed bed operations. Generally, the larger the particle diameter, the cheaper the catalyst. But this is usually not a significant factor for the process designer. More important are uniform fluid flow, pressure drop, and diffusional effects.

Uniform fluid flow through the bed is desirable for good utilization of the catalyst and control of the process. To avoid channeling, the bed is packed as evenly as possible. A rule of thumb dictates that the reactor to particle diameter ratio should be from five to ten, with the reactor length at least 50–100 times the particle diameter.[19] This ensures that the flow is turbulent, uniform, and approximates plug flow. For most commercial reactors these criteria are met. Only in the narrow-tube reactors found in highly endo- or exothermic processes is there any concern.

Pressure drop is a consequence of flow through a packed bed.[20] An extremely high ΔP requires energy-consuming compression and produces undesirable pressure gradients in the bed. Bed voidage is important, and shapes that increase this factor lower ΔP. Figure 1.3 illustrates these principles with pressure drop decreasing in the order spheres > pellets > extrudates > rings > stars or lobes. Particle diameter has a much more

TABLE 1.2. Common Catalyst Particles

Type	Characteristics
Pellets	Made in high-pressure press Shape: cylindrical, very uniform, rings Size: 2–10 mm diam Use: packed, tubular reactors
Extrudates	Squeezed through holes Shape: irregular lengths circular, star or lobe cross section Use: packed, tubular reactors, ebulating beds
Spheres	Made by aging liquid drops Size: 1–20 mm Use: packed tubular reactors, moving beds
Granules	Fusing and crushing, particle granulation Size: 8–14 to 2–4 mesh Use: packed tubular reactors
Flakes	Powder encapsulated in wax Use: liquid phase reactors
Powders	Spray-dried hydrogels Size: <100 μm Use: fluidized reactors, slurry reactors

dramatic result. It would appear that the larger the particles the better. This is true if diffusion is not a problem. If it is a problem, larger particles lead to lower conversions,[21] so some compromise is necessary.

Another parameter of concern to the process designer is the mechanical crushing strength of the particle. If the particle fractures under the weight of the bed or the force of the fluid passing through it, then small fines lodge in interstices between larger particles, causing plugging, uneven flow, hot spots, and pressure drop. Fortunately, crushing strength is fairly independent of particle size.[22] Other factors during preparation and formulation are, however, critical. These are treated in Chapter 6.

We now turn to the question of texture within the particle, i.e., surface area, pore shape, and size distribution. Particles are formulated by agglomerating microparticles produced during a precipitation phase, as shown in Fig. 1.4. Approximately 100 μm in size, these microparticles consist of a complex porous solid. Pores typically range from 1.5 to 15 nm in radius. Formerly termed micropores, these channels are now called mesopores. The name micropores is reserved for those less than 1.5 nm in radius, usually found in zeolites.

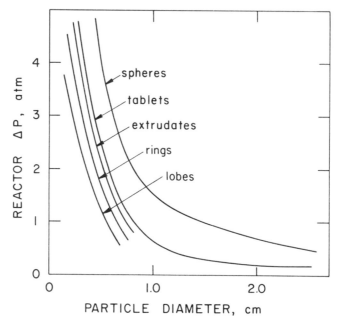

Figure 1.3. Typical reactor pressure drop for catalyst particles.

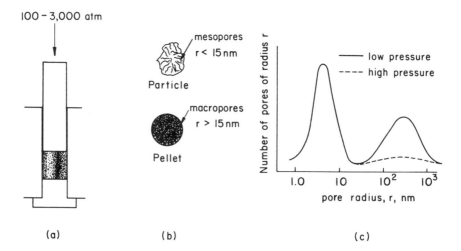

Figure 1.4. Pore structure development in catalyst pellets. (a) Pellet production. (b) Origin of pores. (c) Pore size distribution.

TABLE 1.3. Texture Parameters

Parameter	Nomenclature	Units
Surface area	S_g	$m^2\,g^{-1}$
Pore volume	V_g	$cm^3\,g^{-1}$
Pellet density	d_p	$g(cm^3$ of pellet$)^{-1}$
Porosity	$\theta = V_g d_p$	—
Average pore radius	$r_e = (2V_g/S_g) \times 10^3$	nm
Pore size distribution	PSD	—

During formulation, microparticles are compressed, and grain boundaries flow together to produce rigid, strong particles. Interstices between the microparticles create a macropore structure with radii greater than 1.5 nm. Measurement of pore size distribution shows typical bimodal shapes. Most of the surface area is found in the mesopores, with the combined meso- and macropores amounting to approximately 95% of the total area. Important terms and units used to characterize texture are given in Table 1.3. Measurement of these parameters is discussed in detail in Chapter 7.

1.4. STEPS IN THE CATALYTIC REACTION

The catalytic mechanism extends beyond the surface to involve physical diffusion to and inside the particle. Combining these leads to the steps shown in Fig. 1.5.

1.4.4. External Diffusion

First the reacting molecule, A, diffuses to the external surface of the particle. Motion of A through the fluid outside the particle is governed by external or bulk diffusion. The reader should consult standard references for additional discussion.[21] Useful correlations have been found between the mass transfer factor, j_D, and the dimensionless particle Reynolds number:

$$j_D = 0.57(N_{Re})^{-0.41}, \qquad 50 < N_{Re} < 1000 \tag{1.7}$$

where $N_{Re} = (2R)G/\mu$, R is the particle radius (cm), G is the linear mass velocity (g s^{-1}), and μ is the viscosity (poise).

The mass transfer factor is given by

$$j_D = \frac{k_g \rho}{G}\left(\frac{\mu}{\rho D}\right)^{2/3} \tag{1.8}$$

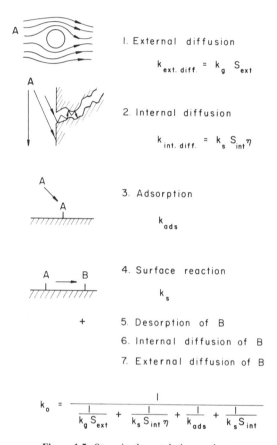

Figure 1.5. Steps in the catalytic reactions.

where k_g is the mass transfer coefficient (cm s^{-1}), ρ is the fluid density (g cm^{-3}), and D_B is the bulk diffusivity of A in the fluid (cm^2 s^{-1}).

This correlation gives k_g with a fair degree of accuracy. Diffusion to the surface is a first-order rate function with a rate constant

$$k_{\text{ext. diff}} = k_g S_{\text{ext}} \tag{1.9}$$

where S_{ext} is the geometrical external surface of the particle per unit particle volume. Equation (1.9) is strictly only valid if the reaction at the surface is very rapid, but it is close enough for most estimates.

Equation (1.7) is good only for particle Reynolds numbers above 50. This suffices for most industrial reactors. Problems arise with laboratory reactors where values of N_{Re} are extremely low. In such cases it is advisable to use the expression[21]

$$j_D = 0.84(N_{Re})^{-0.51} \tag{1.10}$$

Equation (1.10) shows which parameters decrease the rate of diffusion. Increasing the velocity and diffusivity and decreasing particle diameter, density, and viscosity result in an increase in the external diffusion rate. In practice, only adjustments in linear velocity and particle diameter are possible.

1.4.2. Internal Pore Diffusion

The second step in the mechanism is diffusion into the pores leading to the reacting surface sites. Resistance to this diffusion is through collisions either with other molecules (bulk diffusion) or with the walls of the pore (Knudsen diffusion). Satterfield has described methods for calculating bulk and Knudsen diffusivities, D_B and D_K, respectively.[21] It is important to remember that

$$D_B \propto \frac{T^{3/2}}{P_T} \tag{1.11}$$

with T the temperature, P_T the total pressure, and

$$D_K \propto T^{1/2} r_p \tag{1.12}$$

where r_p is the radius of the pore.

Usually D_B and D_K are combined

$$\bar{D} = \frac{D_B D_K}{D_B + D_K} \tag{1.13}$$

and corrected for particle porosity, θ, and tortuosity of the pores, τ,

$$D_{eff} = \frac{\bar{D}\theta}{\tau} \tag{1.14}$$

Porosity is measurable, but τ is difficult to characterize. As an approximation,

$$\tau = 1/\theta \tag{1.15}$$

is useful. If θ is not known, sufficient accuracy is achieved by the approximation $\theta = 0.5$. Thus

$$D_{\text{eff}} = (0.25)\bar{D} \qquad (1.16)$$

is good enough!

For simple nth-order irreversible reactions, models for diffusion-reaction lead to the relationship, shown in Fig. 1.6, between the effectiveness factor, η, and the Thiele modulus

$$\phi = R\left\{\frac{k_s(S_g d_p)[(n+1)/2]}{D_{\text{ext}}} C_A^{n-1}\right\}^{1/2} \qquad (1.17)$$

where k_s is the areal rate constant (per unit surface area) and d_p is the pellet density.

Combining these factors gives a rate constant for internal diffusion

$$k_{\text{int diff}} = k_s S_g d_p \eta \qquad (1.18)$$

For cases of extreme pore diffusion limitations, η is low and

$$\eta = 1/\phi \qquad (1.19)$$

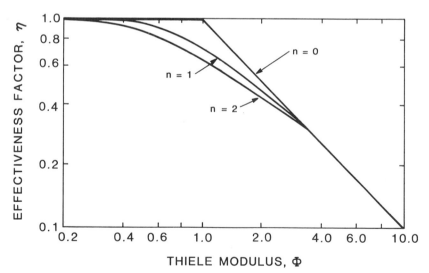

Figure 1.6. Effectiveness factor of an nth-order reaction.

giving

$$k_{\text{int diff}} = \frac{(k_s S_g d_p D_{\text{eff}})^{1/2}}{R[(n+1)/2]^{1/2} C_A^{(n-1)/2}} \tag{1.20}$$

Rates of pore diffusion-controlled reactions can be increased by decreasing the particle radius or increasing the diffusivity. The latter can only be done by increasing the pore radius (without decreasing θ or S_g).

Another important conclusion results from equation (1.20). For the simple-order rate equation

$$\text{rate} = k_{\text{int diff}} C_A^n \tag{1.21}$$

substitution from equation (1.20) gives

$$\text{rate} \propto k_s^{1/2} C_A^{[(n+1)/2]} \tag{1.22}$$

Extreme pore diffusion limitation leads to one half the normal activation energy and moves the order closer to unity. These facts will be useful when we discuss diagnostics.

1.4.3. Adsorption

Having made its way to the interior surface of the porous particle, molecule A is now ready for the first chemical step, adsorption on the surface. In catalysis, adsorption is almost always chemisorption. Chemisorption results from chemical bonds between the molecule (adsorbate) and the solid surface (adsorbent). It is therefore very specific,[23] and receptive sites for chemisorption must exist. Physical adsorption comes from general van der Waals forces, which are physical in origin, weaker than chemisorption, and not specific. Chemisorption stops when a monolayer of adsorbed molecules is formed. It is activated with energies around 10 kcal mole^{-1}, is exothermic with enthalpy changes of -15 to -40 kcal mole^{-1}, is slowly reversible or even irreversible, and is the key step in activation of reaction intermediates.

The rate of chemisorption is governed by the frequency of collisions with the surface and the probability of "sticking" with chemical bond formation. The former is a physical phenomenon, dependent on temperature and pressure. For example, at one atmosphere pressure and 25°C, 3×10^{23} molecules strike each square centimeter of surface each second. If all "stick," the surface is covered in 3×10^{-9} sec.[26] The probability of chemical bonding is exponentially proportional to the enthalpy change, ΔH_a, and activation energy E_a, so the stronger the bond formed the faster the rate and the longer the adsorbed species remains on the surface. Current models for chemisorption include geometric (or ensemble) and electronic (ligand) effects in which a given molecule finds the right symmetry or orbitals at the surface.[23] Understanding these features is important in design of new or improved catalysts and will be treated in Chapter 4.

Also part of the adsorption step is surface migration or diffusion. Although we visualize the adsorptive process as chemical bonding with a surface site, the adsorbed molecule does not stay at that site but "hops" around from one site to another. Activation energy for this process is approximately 0.3–0.5 of ΔH_a. For example, with $\Delta H_a = -30\,\text{kcal mole}^{-1}$ (a typical value) the molecule makes 5×10^{14} "hops" per second![25] The surface is truly a dynamic system with molecules arriving, leaving, and migrating constantly. It is within this framework that we consider the next step of the catalytic mechanism.

1.4.4. Surface Reaction

In this sea of constant movement, two adsorbed molecules destined to react must come together. For instance, the N_S species of mechanism (1.2) finds itself in surface contact with H_S. If surface geometry and energetics are appropriate then a surface reaction occurs. Thus

$$
\begin{array}{ccc}
 & & \text{H} \\
 & & | \\
\text{N} \quad \text{H} & & \text{N} \\
| \quad\ | & & | \quad\ | \\
\text{S} + \text{S} & \rightarrow & \text{S} + \text{S}
\end{array}
\qquad (1.23)
$$

If the N_S species is held too strongly it will either be too immobile to find H_S or disinclined to break its bonds with S and react. If held too weakly, it will desorb before reacting and the surface concentration of N_S is low. Thus we find the situation shown in Fig. 1.7, in which the rate of NH_3 synthesis over different metals passes through a maximum when plotted against the enthalpy of nitrogen adsorption.[26]

Curves of this shape are often found when correlating catalytic rates against some measure of adsorption strength and are called "volcano" curves.

Formulating rate equations from rate-determining steps such as (1.23) is indeed a formidable task. The reader is referred to standard texts on this topic.[27,28] For our purpose in appreciating the role of catalytic steps, it is sufficient to note that most success is achieved with very simple models. By

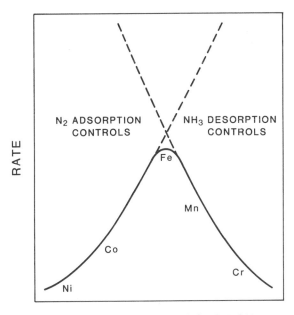

Figure 1.7. Volcano curve for ammonia synthesis.

assuming that all sites are homogeneous (same activation and adsorption energy), immobile, and noninteracting, surface concentrations from the Langmuir isotherm may be used to yield rate equations that fit experimental data as well as any others. The fact that we know that surface sites do not generally follow these criteria need not deter us if we suspect that only a narrow segment of the spectrum of sites is involved in the reaction.[28]

If reaction (1.23) is the rate-determining step (RDS),

$$\text{rate} = k_s \theta_N \theta_H \qquad (1.24)$$

where θ_N and θ_H are the fraction of surface sites occupied by N_S and H_S. From Langmuir isotherms[29]

$$\text{rate} = \frac{k_s K_H^{1/2} K_N^{1/2} P_H^{1/2} P_N^{1/2}}{(1 + K_H^{1/2} P_H^{1/2} + K_N^{1/2} P_N^{1/2})^2} \qquad (1.25)$$

where K_H, K_N are equilibrium constants (or adsorption coefficients) for dissociative adsorption of H_2 and N_2; P_H, P_N are partial pressures of H_2

and N_2. The 1/2 power comes from dissociation of hydrogen and nitrogen molecules. We often make simplifying assumptions where justified, e.g.,

$$(1 + K_H^{1/2} P_H^{1/2}) \ll K_N^{1/2} P_N^{1/2} \tag{1.26}$$

which gives

$$\text{rate} = k_s K_H^{1/2} K_N^{-1/2} P_H^{1/2} P_N^{-1/2} \tag{1.27}$$

or more generally, the Power Rate Law

$$\text{rate} = k_s^q P_H^n P_N^m \tag{1.28}$$

This demonstration is given as an example only. The fact that (1.27) does not fit experimental data is taken as evidence that reaction (1.23) is not the rate-determining step.[30]

Rate equations such as these, based on proposed mechanisms, can become very complex. Often many alternative mechanisms fit equally well. If confirmed, a given rate equation and its associated mechanism serve as a guide only. Other techniques are always required to establish the validity of the surface mechanism.

1.4.5. Product Steps

Steps (5), (6), and (7) in Fig. 1.5 are the reverse of (3), (2), and (1). Although diffusion out is not likely to be different from diffusion in, desorption of a product could be rate determining and must be handled accordingly.

1.4.6. Global Rates

If steps (1)-(4) are all first order and are combined as a series of sequential events, then a first-order reaction with a "global" rate constant k_0 describes the entire process. It can be shown that[31]

$$k_0 = \cfrac{1}{\cfrac{1}{k_g S_{\text{ext}}} + \cfrac{1}{k_s S_g d_p \eta} + \cfrac{1}{k_{\text{ads}}} + \cfrac{1}{k_s}} \tag{1.29}$$

If a given step is much slower than the others (i.e., rate determining), that appropriate term predominates in the global rate constant. If there is more than one slow step, or if the surface reaction is not first order, then a global rate expression is too complicated and cannot be solved analytically.

1.5. DIAGNOSIS OF CATALYTIC FUNCTION

In order to understand and modify the functions of a catalyst in a process, it is necessary to determine whether or not rates are determined by physical or chemical steps. Responses to process parameters and catalyst adjustments are different for the two regimes. Diffusional resistance, in particular, causes unexpected complications. We have seen how low effectiveness factors decrease conversion and disguise kinetics, but selectivity also can be decreased.[32] In addition, poisoning of pore mouth sites in conjunction with low diffusion results in a much more rapid activity decline than otherwise.

Problems with diffusion-controlled processes are eased by changing catalyst properties such as pore size distribution and particle size or process conditions such as linear velocity. Chemical difficulties are treated through modification of the active components of the catalyst.

The remainder of this chapter deals with determining the dominant regime, physical or chemical, prevailing under a given set of conditions.

1.5.1. Interactions between Regimes

Figure 1.8 was designed to demonstrate interactions between physical and chemical rates and to serve as a "first pass" for a process diagnosis. It is calculated for vapor-phase conditions in the temperature range 300–500°C, molecular masses from 2 to 500, and "average" values of D_{eff}. As a teaching tool it serves its purpose without further justification; as a diagnostic device it is useful only within these approximations. However, errors due to deviations from these assumptions are not large considering the range of values. An order of magnitude error in the regime of strong internal diffusion is not serious if the objective is to decide whether further checking is necessary.

Procedures outlined in previous sections were used to compute the curves in Fig. 1.8. Three variables are shown. The abscissa is the chemical rate applicable if all the surface is available. It is the chemical rate built into the catalyst. The ordinate represents values of an index that combines all diffusional tendencies of the particle. Moving in the direction of more diffusion control decreases this index. For example, higher porosity increases D_{eff} (providing it is not accomplished by decreasing the pore radius), increasing total pressure decreases D_B (and thus D_{eff}), and increasing the pellet diameter, D_p, lowers both the effectiveness factor and mass transfer coefficient. Parameters on the curves are pellet rates. The ratio between the pellet rate and the intrinsic chemical rate is the effectiveness factor.

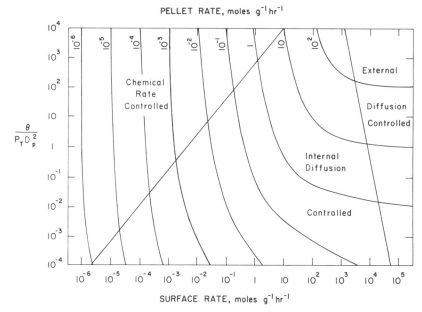

Figure 1.8. Interaction between physical and chemical rate regimes.

If the surface rate is very low, say 10^{-6} moles $g^{-1} hr^{-1}$, and the diffusion index high, >1, then $\eta = 1$ and the surface rate controls. As the chemical rate increases the curves show decreasing effectiveness at increasingly higher diffusion indices. Control passes to internal diffusion and becomes more severe the greater the surface rate. Finally, the surface rate becomes so large that the pellet rate is determined only by diffusion parameters. This is the regime for control by external diffusion. Thus it appears that making a catalyst more active reaches a point of diminishing returns.

Let us demonstrate Fig. 1.8 with an example. A process for steam reforming naphtha is carried out at $H_2O/HC = 10$, 500°C, and 40 atm total pressure. A Ni/Al_2O_3 catalyst, diameter 4 mm, $\theta = 0.6$, was tested in a pilot unit and gave 75% conversion at a GHSV of $5000 \, hr^{-1}$. For pressure drop reasons, the process designer wants to double the size of the particle but keep all conditions the same. Will this change the conversion?

The easiest parameter available is the diffusion index

$$\frac{\theta}{P_T D_p^2} = \frac{(0.6)}{(40)(0.04)^2} = 9.38 \times 10^{-2} \approx 10^{-1} \qquad (1.30)$$

Unfortunately, the surface rate is not available. This is usually the case unless extensive research and development data are known. Working with commerical catalysts frequently precludes fundamental information and we are forced to use pilot unit or even plant data similar to this example.

To calculate the pellet rate, proceed as follows. For an "average" naphtha molecule, C_7H_{16}, vapor density at 25°C, $\rho = 4.46 \times 10^{-3}$ g cm^{-3}, molecular weight, $M = 100$, and pellet density, $d_p = 2.5$ g cm^{-3}

$$\text{rate} = \frac{(X_{HC})(\text{GHSV})(\rho)}{(M)(d_p)}$$

$$= \frac{(0.75)(5000)(4.4 \times 10^{-3})}{(100)(2.5)} = 7 \times 10^{-2} \qquad (1.31)$$

Intersection of these two lines in Fig. 1.8 is at a point where the chemical rate is about 7×10^{-1} moles g^{-1} hr^{-1}, i.e., $\eta = 0.1$. Clearly this is well within the internal diffusion regime, even with all the simplifying assumptions. Doubling the pellet size will decrease conversion to half its value. Even though this analysis may be tenuous, it certainly justifies further tests with 8-mm particles.

1.5.2. Diagnostic Tests

Although this "back of an envelope" calculation is sufficient for many purposes, other indications may be used, depending on the amount of information available. Absence of such data is itself a valuable indication of where to begin experimental verification.

When faced with the diagnosis of process behavior, it is sometimes useful to answer the following questions:

1. What is the apparent order of the reaction? Reaction orders tend towards first order as the process becomes diffusion controlled.[21] First order should be regarded as potentially suspicious. Reaction orders other than one probably indicate chemical control.

2. What is the activation energy? Most chemical activation energies found in catalysis are in the range 25–50 kcal mole^{-1}. Pore diffusion resistance decreases this by a factor of 2.[21] Values from 10 to 20 kcal mole^{-1} should be regarded as indications of pore-diffusion control. External diffusion follows either a 3/2 or 1/2 order in temperature. This appears to be an activation energy of about 5 kcal mole^{-1}—definitely suspicious of external diffusion.

3. Does the rate depend on particle size? If the rate of reaction under constant conditions is inversely proportional to the radius of the particle, this is a strong indication of pore diffusion resistance. External diffusion also shows size effects, but not nearly so pronounced.

4. Does the rate depend on linear velocity? With external diffusion, the rate at constant space velocity increases with linear velocity, at about 1/2 order. Velocity independence is a good indication of the absence of external diffusion effects.

5. What does the observed rate indicate? Take any conversion data, laboratory or plant, whichever are available. Make calculations as shown in the example. If more precision is needed, use any of the criteria given in standard references.[33]

2

STRUCTURE OF CATALYSTS
What They Contain

2.1. CATALYST ENGINEERING

The ability to engineer desired functions into particles is perhaps the ultimate achievement of the catalyst designer. In many cases, direct and obvious procedures are sufficient; in others, more subtle and innovative measures are called for. Although catalyst designs tend to evolve slowly as process experience accumulates, anticipating future problems will decrease development time.

In this chapter, we examine the architecture or structure of the catalyst; what it contains and why.

2.2. THE BALANCED PARTICLE—A COMPROMISE

Correct formulation of the catalyst is a compromise between good fluid flow, activity, and stability. The relative importance of these factors depends upon the reaction, reactor design, process conditions, and economics. In Fig. 2.1, these are represented as three sides of a triangle to emphasize mutual dependence and interaction.

Good flow distribution and low pressure drop are achieved through proper selection of the shape and size of the particle,[34] mechanical strength during manufacture.[35] These are questions of formulation—making the particle—and must be matched to demands of the process for which the catalyst is intended. Generally, the greater the severity of process conditions (e.g., heavier feedstocks, higher temperatures and pressures, larger space velocities) the greater these demands become. For example, the preferred

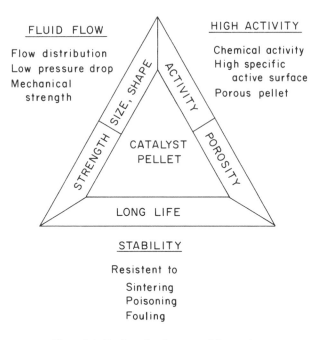

Figure 2.1. Engineering features of the catalyst.

catalyst for hydrotreating, $CoMo/Al_2O_3$, is found in a wide variety of shapes and sizes. Pellets, extrudates, and spheres, 1-10 mm, are all available, depending on whether the feedstock is light naphtha, gas oil, or vacuum residua.[36]

High activity and selectivity is engineered by selecting correct chemical components, using preparational methods to give the required surface area, and formulating the pellet to ensure good access to active sites. Achieving activity alone is not sufficient unless balanced with these other requirements.

Lifetime stability requires resistance to deactivation due to sintering (loss of active surface through crystallite growth), poisoning (elimination of active sites by strong chemisorption), and fouling (blockage of the pores and surface by carbon and debris). Although also sensitive to process conditions, stability is best achieved through addition of catalyst components.

An interesting example that demonstrates similar chemistry for different process conditions is found in the reversible reaction

$$CH_4 + H_2O \rightleftarrows CO + 3H_2 \tag{2.1}$$

Both forward and reverse reactions are catalyzed by the same material, i.e., nickel.[37] However, the state of the nickel is profoundly influenced by process conditions. The forward reaction is steam reforming, an important source of hydrogen for the ammonia and methanol synthesis, iron ore reduction, and petroleum hydrotreating.[38] The reaction is endothermic and equilibrium limited, so that high temperatures (700–1000°C) are needed for high yields. Under these conditions, nickel sinters rapidly. But activity is not important since kinetic rates at high temperatures are sufficient. High dispersions of nickel are not necessary.

Because of the extreme endothermicity, the reaction is heat-transfer limited. Narrow reactor tubes (10–20 cm) ensure high surface to volume ratios and good heat transfer but need sufficient length for economic space velocities ($>10,000 \text{ hr}^{-1}$). Thus pressure drop is a problem. Large particles are called for (Fig. 2.1), but in narrow tubes this causes flow problems if the tube to particle diameter ratio falls below 5–10. Also, at these high temperatures, reaction rates are so fast that the effectiveness factor is low. Some relief is found by using ring-shaped particles, which not only increase bed voidage but also give lower effective diameters.[38] But some degree of compromise is necessary, leading to the use of particles about 2 cm in diameter.

In addition, severe temperatures require thermal stability. Particles must retain their physical and mechanical properties, and such things as phase transitions and fracturing must be avoided. Thus, an important feature in designing steam reforming catalysts is a suitable support giving strength to the particle and stability to the nickel. Many solutions have emerged, e.g., $MgAl_2O_4$- or $CaAl_2O_4$-based systems.

The reverse of reaction (2.1) is methanation. Used to remove residual CO traces from ammonia synthesis feedstocks, it was also developed as an important source of substitute natural gas (SNG) in the synthetic fuels industry.[4] Since this reaction is exothermic, equilibrium yields are better at low temperatures (300–500°C). Thus, high activity is critical. Nickel must be highly dispersed. Preparational methods are required to produce small nickel crystallites. This high metal area must be maintained in the presence of extreme exothermicity, so that sintering must be avoided. This is partially accomplished through proper[39] catalyst design, but process reactor type must also be considered. Recycle, fluidized, and slurry reactors are appropriate.

Questions of lifetime stability are different for the two processes. We have discussed thermal degradation and sintering. Also important are sulfur poisoning and carbon fouling. Sulfur rapidly deactivates nickel sites by adsorption of sulfur atoms.[40] For steam reforming, temperatures are high enough that steam removes the sulfur, and higher levels may be tolerated.[38] With methanation, sulfur poisoning is irreversible and the only protection is to desulfurize the feed.[4]

For both processes, carbon formation is serious.[41] In reforming it originates mostly from CH_4 decomposition, in methanation from CO disproportionation. Preventing nickel from catalyzing these conversions is difficult without affecting the main reactions. The solution is to remove the carbon through reaction with steam before polymerization to graphite occurs. This is done by adding another catalyst (in this case a promoter) to the formulation.[42] Alkalis catalyze carbon removal, the most effective being potassium. The fact that they also lower nickel activity for the main reaction is another compromise that must be accepted.

In this example, the delicate balance between three catalyst factors and process conditions is clearly demonstrated. Other examples will be seen over and over again as we examine successful, proven catalyst developments.

2.3. CATALYST COMPONENTS

Although some catalytic materials are composed of single substances, most catalysts have three types of easily distinguishable components: (1) active components, (2) a support or carrier, and (3) promoters.[43] In Fig. 2.2 we again invoke a triangle to demonstrate mutual dependencies.

2.3.1. Active Components

Active components are responsible for the principal chemical reaction. Selection of the active component is the first step in catalyst design. As knowledge of catalytic mechanisms on various materials advances, methods for selection are becoming more scientific, if perhaps still empirical. These are examined in subsequent chapters.

Historically, it has been convenient to catalog active components according to the type of electrical conductivity (Table 2.1). Table 2.1 is not intended to be exhaustive but to give perspective to the classification. More examples are given in Chapter 4.

The main reason for classifying active components by conductivity type is one of convenience. No relationship between conductivity and catalysis should be assumed. However, both depend on atomic electronic configurations. Each of the three, metals, semiconductors, and insulators, has a theoretical and experimental background useful in deriving catalytic models. With metals, overlapping electronic energy bands promote electron transfer with adsorbing molecules.[44] Redox or charge transfer reactions such as hydrogenation, hydrogenolysis, and oxidation are found. Systematic variation with atomic electron configurations explains trends in adsorption and catalysis. Similarities in groups within the periodic table are rationalized. Orbital considerations, such as type, occupancy, and symmetry, may be

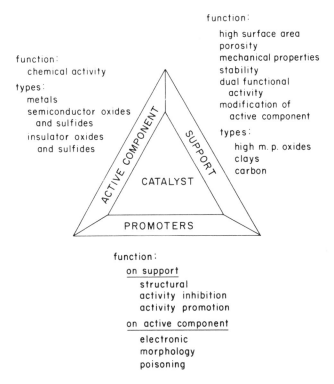

Figure 2.2. Catalyst components.

TABLE 2.1. Classification of Active Components

Class	Conductivity/ Reaction type	Reactions	Examples
Metals	Conductors	Hydrogenation	Fe, Ni, Pt
	Redox	Hydrogenolysis	Pd, Cu, Ag
		Oxidation	
Oxides and sulfides	Semiconductors	Selective	NiO, ZnO, CuO
	Redox	hydrogenation	
		Hydrogenolysis	Cr_2O_3, MoS_2
		Oxidation	
Oxides	Insulators	Polymerization	Al_2O_3, SiO_2, MgO
	Carbonium ions	Isomerization	SiO_2-Al_2O_3
		Cracking	
		Dehydration	Zeolites

invoked to explain and predict differences in, for example, metals or crystal planes.

Semiconducting oxides and sulfides constitute a large class of catalytic materials.[45] Electron donor and acceptor levels provide redox-type activation, but surface configurations are more complicated than with metals. Greater geometric complexity leads to more selective redox reactions, such as partial oxidation, hydrodesulfurization, and denitrogenation. Insulators do not readily promote charge transfer, but surface sites with localized protons are favored.[46] Acidlike in nature, these sites promote carbonium ion mechanisms, resulting in typically acid-catalyzed reactions such as isomerization or cracking.

With these brief introductory remarks, the reader is referred to Chapter 4 for more explicit descriptions of active components.

2.3.2. Support

Supports, or carriers, perform many functions, but most important is maintenance of high surface area for the active component. This is best illustrated with platinum, an important catalytic metal widely used for catalytic reforming[11] and automobile exhaust clean-up.[47] For high activity, platinum crystallites must have the highest surface area possible. Figure 2.3 shows the relationship between dispersion, defined as the fraction of platinum atoms on the surface of the spherical crystallite, and diameter of the sphere. Dispersion decreases very rapidly between 1 and 10 nm. Ideally, platinum crystallites should be as small as possible, but certainly

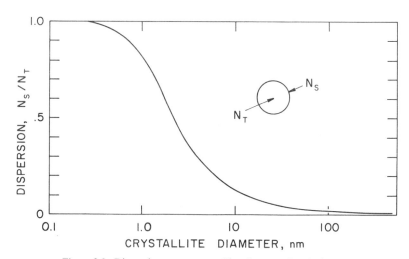

Figure 2.3. Dispersion versus crystallite diameter for platinum.

in the 0.5–5 nm range. Platinum spheres of this size can be made as colloidal platinum black.[48] Attempts to use the colloid as a catalyst at reaction temperatures rapidly leads to sintering or agglomeration of the crystallites through the mechanism shown in Fig. 2.4.

Crystallites contact each other through thermal motion, becoming especially agitated at higher temperatures. Above the "Tamm" temperature $0.5\,T_m$, where T_m is the melting point, bulk metal atoms acquire enough thermal energy to migrate within the crystallite. Even at $0.3\,T_m$ (the Huttig temperature) surface atoms have enough energy to overcome weaker surface crystal forces, diffuse, and form necks as shown in Fig. 2.4. If surface and bulk atoms are mobile enough, two smaller crystallites coalesce to a larger one, thereby decreasing the surface energy. Typical growth is both a thermodynamic and kinetic effect. Patterns showing a dependence on melting point are demonstrated in Figure 2.5.[49]

Platinum has a melting point of 1774°C. Even if prepared as platinum black, 1 h of reaction at 400°C results in 50 nm crystallites, six months in 2000 nm. Such instability is unacceptable.

Supports function as stable surfaces over which the active component is dispersed in such a way that sintering is reduced.[43] The support itself

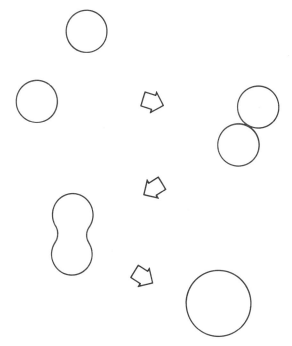

Figure 2.4. Sintering of colloidal metal crystallites.

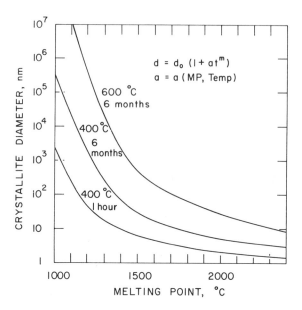

Figure 2.5. Equilibrium size dependence on melting point.[49]

must be secure from thermal growth, which means high melting point—at least higher than that of the active component. High melting points are found in oxides normally considered as ceramic materials. These are listed in Table 2.2.

In principle, these and many more high melting oxides are potential candidates for supports. However, the oxide must be amenable to colloidal preparations yielding high surface areas. If high dispersions are not needed, as in steam reforming, then less demanding preparational methods are sufficient.

High area supports commonly used in catalyst manufacturing are listed in Table 2.3.

The level of loading is important to the role of the support in maintaining dispersion of the active component. Crystallites, even though isolated from each other on the support surface, may sinter. For very small crystallites, sintering occurs through migration over the surface followed by coalescence. Important factors, besides temperature, are crystallite concentration, interaction with the support, and atomic mobility.[50] The effect of concentration is shown in Fig. 2.6.[51]

With Ni/Al_2O_3 catalysts, total nickel area increases with loading. Even up to about 40 wt % nickel the crystallites are sufficiently separated that extensive sintering during reduction does not occur. Above 50%, however,

TABLE 2.2. High Melting Oxides Used as Supports

Type	Oxide	Melting point (°C)
Basic	MgO	3073
	CaO	2853
	Ca_2SiO_4	2407
	BaO	2196
	Ca_3SiO_5	2173
Amphoteric	ThO_2	2323
	ZrO_2	2988
	CeO_2	2873
	Cr_2O_3	2708
	La_2O_3	2588
	α-Al_2O_3	2318
	TiO_2	2113
Neutral	$MgAl_2O_4$	2408
	$MgCr_2O_4$	2300
	$ZnCr_2O_4$	2173
	$ZnAl_2O_4$	2100
	$CaSiO_3$	1813
Acidic	γ-Al_2O_3	2318
	SiO_2	1973
	SiO_2-Al_2O_3	1818

interactions between the crystallites increase, growth occurs, and total nickel area decreases. Activity per unit volume of catalyst passes through a maximum. The exact position of the maximum may be controlled through preparational techniques and the use of additives.

Supported metals have either low (e.g., 0.3% Pt/Al_2O_3, 15% Ni/Al_2O_3) or high loading (70% Ni/Al_2O_3, Fe/Al_2O_3), depending on the process. In the case of high loading, the term "support" is still used, although, from the representation shown in Fig. 2.7, spacer might be more appropriate.

Porosity is necessary for high surface area within the pellet, but pore shape and size distribution are critical secondary factors when diffusion

TABLE 2.3. High Area Supports
Commonly Used

γ-Al_2O_3
SiO_2
C (activated)
Diatomaceous clays
SiO_2-Al_2O_3

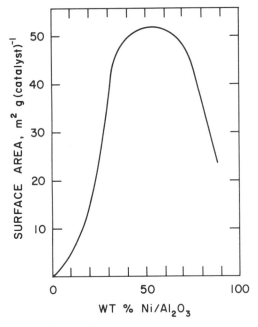

Figure 2.6. Surface area for alumina-supported nickel catalysts.[51]

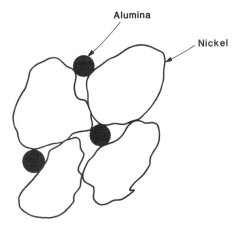

Figure 2.7. High loaded catalyst compositions.

resistance is present.[52] The best supports are those that are easily manipulated to produce optimum texture properties. Alumina and silica are good in this regard.[53,54] Much experience has been accumulated on the hydrothermal treatment of hydrous alumina and silica gels. Catalyst manufacturers put this to good use in producing supports with controlled pore shapes and sizes. Examples are given in Chapter 6.

Mechanical strength and thermal stability of catalyst particles are always of concern to process designers. In some cases it may be the most critical feature. This was emphasized, for example, in steam reforming. Strong pellets with good thermal resistance are required. Catalyst designers use mixed oxides fired at high temperatures to form ceramic compounds. Particles must be preformed and active components added later.

Ideally, support materials should have no catalytic activity leading to undesirable side reactions. This is usually true for high melting oxides fired to give low surface area supports. However, colloidal hydrous oxides are usually acidic in nature. Alumina, for example, is dehydrated during preparation:

$$
\begin{array}{c}
\overset{\displaystyle OH}{\underset{\displaystyle |}{}} \quad \overset{\displaystyle OH}{\underset{\displaystyle |}{}} \qquad\qquad\qquad \overset{\displaystyle O^-}{\underset{\displaystyle |}{}} \\
-O-Al-O-Al \xrightarrow[-H_2O]{} \; O-Al^+-O-Al-O- \qquad (2.2) \\
\underset{\substack{\text{Lewis} \\ \text{acid} \\ \text{site}}}{} \quad \underset{\substack{\text{Basic} \\ \text{site}}}{}
\end{array}
$$

However, there is always sufficient adsorbed water present on the catalyst surface to give

$$
\begin{array}{c}
\overset{\displaystyle H^+}{} \\
\overset{\displaystyle O^-}{\underset{\displaystyle |}{}} \qquad\qquad \overset{\displaystyle \overset{..}{O}H}{\underset{\displaystyle |}{}} \quad \overset{\displaystyle O^-}{\underset{\displaystyle |}{}} \\
-O-Al^+-O-Al-O \xrightarrow{+H_2O} O-Al-O-Al- \qquad (2.3) \\
\underset{\substack{\text{Bronsted} \\ \text{acid} \\ \text{site}}}{}
\end{array}
$$

Bronsted sites, that initiate carbonium ion reactions, and Lewis sites, giving ion radical reactions, coexist, althouth it appears that in practical usage the Bronsted acidity predominates.[55] When using γ-Al$_2$O$_3$ as a support, undesirable side reactions such as cracking, isomerization and "coke" formation always exist. These give unwanted products and lead to catalyst deactivation.

There are situations where support acidity has a positive influence, influencing the main reaction. The support adds dual functionality to the overall catalysis, as best demonstrated with catalytic reforming.[56] The objective in this process is to convert low octane components of naphtha, typically normal paraffins and naphthenes, into high-octane iso-paraffins and aromatics. Low loadings of Pt-type metals on Al_2O_3 are used for this purpose. Metallic Pt dehydrogenates naphthenes to aromatics but cannot isomerize or cyclize normal paraffins. This is accomplished through the acidic function of the support, as shown for n-hexane:

$$n\text{-}C_6 \qquad\qquad i\text{-}C6$$
$$-H \downarrow_{Pt} \qquad\qquad +H \uparrow_{Pt} \qquad\qquad\qquad (2.4)$$
$$n\text{-}C_6^{=} \xrightarrow[Al_2O_3]{H^+} i\text{-}C_6^{=}$$

Neither Pt or Al_2O_3 promotes paraffin isomerization, but $Al_2O_3^-$ acid sites isomerize normal olefins. In a three step sequence, n-C_6 is dehydrogenated by Pt to n-$C_6^{=}$, which migrates to the Al_2O_3 and isomerizes to i-$C_6^{=}$, then hydrogenated by the Pt to i-C_6. Platinum and Al_2O_3 must be in intimate contact and the Al_2O_3 have sufficient acidity. Chlorine is added during activation to generate the needed acidity.[56] Dual functionality is indeed an elegant demonstration of catalyst design. Another important example is hydrocracking.[57]

Catalytic chemists have long suspected that the properties of the active component may be affected by contact with the support. Indeed, as we shall see, there are so many ways in which supports exert influences that it is not surprising that these were noted. It was not until recently that researchers were able to unravel some of the these effects with more sophisticated characterization methods. For example, measured areal rates of ethane hydrogenolysis showed the patterns in Table 2.4.[58]

The most direct influence of the support is on dispersion and morphology.[59] It is well known that nickel is better dispersed on SiO_2 than Al_2O_3, and the shape of crystallites may also be affected. Data in Table 2.4 do not address the question of crystallite size dependence, which has a much firmer foundation.[60] Surface contamination is always present. Commercial reagents used in support preparation may contain impurities or provide ions which remain with the oxide. Subsequent deposition of active components may incorporate these ions as poisons. Acidic sites vary on different supports and during initial exposure to reagents produce carbonaceous deposits which in turn deactivate nickel[61] sites. A possibility that cannot be overlooked is spillover, a phenomena in which a reactive

TABLE 2.4. Ethane Hydrogenolysis with
Different Supports[a]

Support	Rate [moles m^{-2} hr^{-1}(metal) × 10^6]	
	10% Ni	10% Co
SiO$_2$	151	230
Al$_2$O$_3$	57	185
SiO-Al$_2$O$_3$	7	10

[a] Reference 58.

species is generated on a metal site and then migrates over the support to other sites where further reaction occurs.[62]

The most favored explanation is electron transfer, in which electrons are moved from or to the active component due to electron donating or accepting sites on the support. Electron density in the active component changes much like in alloys. This effect was observed much more dramatically in the case of reducible supports such as TiO$_2$, which provide a large enough concentration of altervalent ions for electron transfer. However, the effect has not been conclusively proven.[63]

2.3.3. Promoters

A promoter is some third agent which when added, often in small amounts, results in desirable activity, selectivity or stability effects. It is perhaps in this regard that catalysis still deserves the term "black art" and the catalyst designer "alchemist". Promoters are like spices in cooking, a pinch is added here and there because we know it makes the recipe taste better. Indeed, as in cooking, many promotional effects are discovered accidentally. We then glorify them with "scientific" explanations. There is much work to be done in this part of catalysis.

Promoters are designed to assist either the support or the active component. One important example of support promotion is control of stability. Support oxides may occur in several different phases, some undesirable. With Al$_2$O$_3$, for example, the preferred phase is γ-Al$_2$O$_3$.[64] A defect spinel, this phase has high surface area, a certain degree of acidity, and forms solid solutions with transition oxides such as NiO and CoO. When heated, γ-Al$_2$O$_3$ transforms into α-Al$_2$O$_3$, which has an hexagonal structure and low surface area. The transition begins measurably at about 900°C, a temperature not usually encountered during process conditions but possible during catalyst regeneration. Furthermore, the phase change occurs slowly at lower temperatures, which for anticipated lifetimes (3+ years) could result in significant changes. Incorporation of small amounts of SiO_2 or ZrO_2 into the γ-Al_2O_3 (only 1-2%) moves the α-Al_2O_3 phase transition to higher temperatures.[65] The support is then adequately protected against major upset and long term changes.

Most often, promoters are added to supports in order to inhibit undesirable activity, such as coke formation. Coking originates from cracking on Bronsted sites followed by acid-catalyzed polymerization to give $(CH_x)_n$ species that cover surface sites and ultimately block pores. Removal of the coke by burning may itself lead to activity loss due to sintering. Acidic cracking sites are neutralized by bases, most effectively alkalis.[66] Potassium, added as potash during preparation, lowers the coking tendencies of alumina supports. Sodium is used also but appears to facilitate sintering and is more mobile, often poisoning the active component.

When dual-functional activity is needed, as in catalytic reforming, extra acidity is achieved by adding chloride ions to the surface.[56] Since chlorine is removed during regeneration, it must be replenished during reactivation or chlorine compounds must be added to the feed during operation.

TABLE 2.5. Examples of Promoters in Major Processes

Catalyst	Promoter	Function
Al_2O_3 Support and catalyst	SiO_2, ZrO, P K_2O HCl MgO	Improves thermal stability Poisons coking sites Increases acidity Retards sintering of active component
SiO_2-Al_2O_3 Cracking catalyst and matrix	Pt	Increases CO oxidation
Zeolites Cracking catalyst	Rare earth ions Pd	Increases acidity and thermal stability Increases hydrogenation
Pt/Al_2O_3 Catalytic reforming	Re	Decreases hydrogenolysis and sintering
MoO_3/Al2O3 Hydrotreating	Ni, Co P, B	Increases hydrogenolysis of C-S and C-N Increases MoO_3 dispersion
Ni/ceramic supports steam reforming	K	Increases carbon removal
Cu-ZnO-Al_2O_3 Low temperature shift	ZnO	Decreases Cu sintering

Promotion of the active component may be either structural or electronic.[67] For metals, there are many well documented examples. Ammonia synthesis catalysts consist of iron doubly promoted with alumina and potassium.[2] It was long thought that the role of alumina was to prevent iron sintering upon reduction and that potassium poisoned acid sites inadvertently introduced by the alumina.[68] Now, there are indications the alumina promotes the iron by stabilizing [111] surface planes which are 500 times more active than other planes.[67] The potassium also doubles as an electronic modifier by donating electrons from the ionized state.[14] These improve π-bonding of nitrogen, leading to faster dissociation and higher rates. Similar effects have been reported for hydrogenation of carbon monoxide on nickel.[67]

Another successful promotion is in catalytic reforming, where addition of rhenium to platinum leads to vastly improved performance, due to decreased hydrogenolysis of the hydrocarbons.[11] The rhenium functions perhaps as an electronic promoter, but its exact role is still being debated. Chemically, the rhenium preferentially bonds with low coordination sites on crystal plane corners, edges and steps. Since these atoms appear to be involved with the hydrogenolysis, coke-forming reactions are suppressed.[67]

With oxidic active components, promoter interactions can also be structural or electronic. A good example is $CoMo/Al_2O_3$, discussed in detail in the next section.

Table 2.5 lists common promoters and their mode of action. In principle, any additive which enhances or inhibits catalytic functions can be classed as a promoter.

4. EXAMPLE—HDS CATALYSTS

Here we demonstrate the unique role of all three components in the case of $CoMo/Al_2O_3$ catalysts, used in hydrodesulfurization processes. Many different types of organic sulfur compounds are involved and the reaction follows a series[15]

$$R-S \xrightarrow[-H_2S]{+H_2} R^= \xrightarrow{+H_2} R \tag{2.5}$$

Hydrogenolysis of C–S bonds is followed by hydrogenation. Petroleum fractions are desulfurized for many reasons; to protect catalysts, to improve product quality, and to prevent environmental pollution.[69] Feedstocks vary from light naphthas to heavy residua and require reactors of increasing complexity, from fixed to trickle bed reactors and even ebullating processes. Catalyst particles are available in a wide range of sizes and shapes to fill these needs.[70]

The fundamental chemistry, however, remains the same. The preferred support is γ-Al_2O_3 prepared by precipitation and preformed into particles with designed pore size, shapes, and distribution. Details are given in Chapter 6. High surface area ($\sim$250 m^2 g^{-1}) is achieved during preparation. Small amounts ($\sim$1%) of SiO_2 are added to stabilize the high area phase.

The active component is a sulfided molybdenum compound promoted with cobalt. Cobalt and molybdenum salts are deposited onto the alumina and calcined to a mixture of MoO_3, CoO (Co_3O_4) and $CoAl_2O_4$.[71] Molybdena is the precursor of the active component. Cobalt aluminate originates from solid state reactions between CoO and Al_2O_4. Surface studies reveal that this compound forms a substrate on which a monolayer of MoO_3-type structures deposits. Some cobalt is incorporated within this monolayer, as shown in Fig. 2.8.[72]

The catalyst is activated through sulfiding, either by high-sulfur feed or treatment. During this process the MoO_3 layer breaks up into microcrystals of MoS_2, into which small amounts of cobalt ions are incorporated,

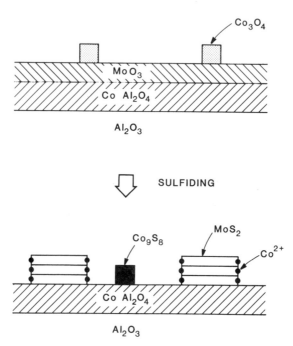

Figure 2.8. Surface structure of CoMo/Al_2O_3 catalysts.

effectively "doping" the electronic structure and producing a "CoMoS" active site.[73]

The promotion effect of the CoMoS site appears to occur through increased hydrogen activation, which facilitates removal of sulfur atoms after cleavage of C–S bonds on exposed molybdenum ions.[74] Cobalt incorporated into the support may also act as a structural promoter by enhancing dispersion of the sulfided species.[75]

With heavier feeds, potassium is added to poison acid sites and inhibit coking. Phosphorous is also used, presumably to inhibit acidity but also to promote dispersion of the molybdenum compounds.[76]

One additional feature occurs when this catalyst is used with extremely heavy residua feedstocks. With proper preparational techniques, bimodal pore distributions are produced in which constrictions at the openings inhibit diffusion of large asphaltenic, coke-producing molecules, whereas smaller sulfur compounds have easy access.[77] This effectively reduces excessive coking which otherwise occurs.[69]

In this example, we see how the catalyst designer "tailors" the components of the catalyst to meet process needs. This is not usually a spontaneous, creative event but one that evolves with experience. How this occurs and how catalysts are developed is considered in the next chapter.

CATALYST DEVELOPMENT
How They Evolve

3.1. CATALYST DEVELOPMENT

Catalyst development is a process of evolution. Although cases differ, successful process implementation follows a series of logical steps. Each has well-defined objectives that require talents from established branches of science and engineering. Commercialization may involve many passes through these steps, but every proven catalyst development has, in some fashion, progressed in this manner. In this chapter we examine elements of catalyst development and offer guidelines for success.

A "flow sheet" for catalyst development is shown in Fig. 3.1. Readers will recognize their own roles in this sequence. When placed in proper perspective, contributions from others are appreciated in what must be, by necessity, a fully integrated effort.

3.2. PROCESS NEED

Every new catalyst results from a process need. Only catalyst salesmen seek new applications for existing catalysts. The author was once told by a colleague that he had invented a new catalyst and "all I need is a reaction." Humerous as this sounds, it underlines a very important point. Catalyst technology responds to needs of society and the marketplace. Economic, social, and political forces have all left their marks on today's industries.

Either an entirely new process or some modification of an existing process is called for. Here we classify a process as new if it involves different

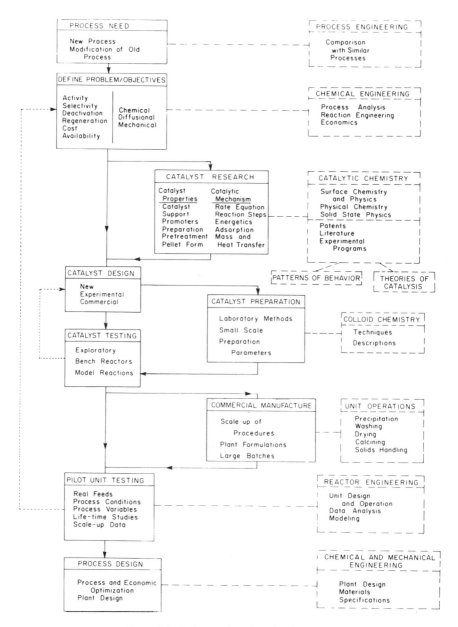

Figure 3.1. Pathway of catalyst development.

feedstocks, products, processing routes, or conditions from those normally used. Direct liquefaction of coal is a recent example; low-density polyethylene is another.

Modification of an existing process involves a change either in conversion, selectivity, lifetime, or operating conditions. The need may come from changes in economics, feedstocks, products and energy supplies or from problems encountered in established units. For example, steam reforming, as practiced until the 1960s, was a fairly low-pressure process. In order to take advantage of more economical centrifugal compressors, it was desirable to raise pressure in the reformer.[1] Higher pressures required more active and robust catalysts. It is interesting that the solution to this problem was to increase silica in the support, thereby generating a new problem. At higher steam pressures, silica, which is volatile, leaves the catalyst, and deposits on lower-temperature equipment downstream. Thus a second-generation solution was needed.[49]

Whether new or existing, it is desirable at this point to make a detailed analysis of similar processes. This helps to focus upon potential problems and also define the direction of the development. For example, in direct coal liquefaction, comparison with hydrotreating of heavy petroleum fractions was useful to establish similarities and differences.[78]

Detailed comparisons are time consuming and tedious, but well worth the effort. It is wise to follow a systematic procedure such as the "check list" in Appendix 4. Only in this way is it certain that all relevant data are included and nothing overlooked. If conscientiously carried out, the exercise is not only informative but often generates creative and innovative ideas.

3.3. DEFINE PROBLEM AND OBJECTIVES

Catalyst problems vary from case to case. Increased activity or improved selectivity may be called for, or decreased deactivation. Perhaps it is a question of devising satisfactory regeneration procedures. In some cases cost and availability of catalyst components may be the factor. Analyses of similar processes are helpful in establishing reasonable levels and limits. Just how much activity is required? What selectivity is sufficient? What is the minimum regeneration cycle?

Next, we must define the precise objectives necessary to solve these problems, and this can be very critical. For example, activity, selectivity, and deactivation problems may exist because of either chemical or diffusional factors. Which is it? The answer determines the route of future development. If chemical properties are to be adjusted, then modification of surface composition is necessary. On the other hand, correction for diffusional resistance requires changes in pore structure and particle formulation. The methods for diffusional diagnosis given in Chapter 1 should be used here, even if only limited data on comparable processes are available.

Equally important are problems in particle formulation. Particle shape, size, strength, and stability must be considered. Often these features coexist with chemical and diffusional difficulties.[34]

These first two steps require the talents of process engineers. Knowledge of process design, reaction engineering, and economics is invaluable for proper diagnosis and analysis.[79,80,81]

3.4. CATALYST RESEARCH

In some cases, the problems and objectives are so obvious that the path leads directly to catalyst design, for example, decreasing particle size for higher conversion or modifying formulation for extra strength. However, most situations divert to some extent into an area of research.

Catalysis research is a vast arena, utilizing specialties of many disciplines, from surface physicists and theoretical chemists on the one hand to chemical engineers on the other.[82] Innovative and outstanding work is being done at academic and industrial laboratories all over the world. High-technology computers and equipment enable us to calculate and observe the behavior of real molecules on well-characterized ideal surfaces.[83] Preparation of powders is precisely controlled, properties exactly measured, and kinetic experiments carried out with hitherto unknown confidence. Modeling of catalyst particles and beds has reached a stage where complex reaction schemes and deactivation regimes are handled with ease.[84] The catalysis literature, documented in Appendix 1, has exploded with an influx of books and articles, attesting to the high level of research in progress.

The catalyst designer, seeking solutions to his problems, should not overlook any source of information. Not only the open literature but also patents and reports of experimental programs should be searched for relevant data. This may be enough to proceed with catalyst design, in which case the effort is justified. If nothing more, gaps in the literature indicate experimental research worth doing.

For designing and developing catalysts, the focal point of all research, the pay-off, is correlation between catalyst properties on the one hand and mechanisms on the other. Choice of the active components, type of support and promotion, method of preparation, pretreatment, and formulation must be related to mechanism, kinetics, adsorption, or mass transfer. Often data are incomplete or speculative. Correlations may be empirical but prove extremely useful. Utilization rather than analysis is often more fruitful; success its own reward. It is not surprising that commercial solutions in catalysis almost always outdistance scientific understanding. And yet it is only through the pursuit of precise and accurate models that we advance one more step and facilitate success in future endeavors. The research phase ultimately reaches a stage where expediency or financial limitations dictate that the next step be taken with the information at hand. The better prepared we are to interpret this information, the more successful will be our designs.

3.5. CATALYST DESIGN

Catalyst design involves selecting candidates for further testing and optimization. These may be very close to the final form if based on previously proven catalysts. For novel designs, however, considerable adjustment lies ahead.

To the uninitiated, designing a catalyst is a mysterious process. However, familiarization with the chemistry of the process, and especially the mechanism, narrows the choice of possible materials. By comparison with similar reactions, by use of correlations and patterns of behavior, and from current research, it is possible to arrive at fairly good estimates as a starting point.

Systematic, scientific methods for catalyst design have been pursued by researchers for many years. A recent book gives many examples[85] and more details and guidance are found in Chapter 5.

Candidate catalysts from the design may be new or novel compositions, giving the designer a proprietary position. However, commercialization takes longer since there are many steps remaining before the worthiness of the invention is demonstrated. To solve an immediate process problem, the best solution is a design already proven or commercialized. Having survived manufacturing scale-up and implementation, such a catalyst is much more readily available for dependable testing.

Another possibility is a development catalyst. These are catalysts, usually under development by catalyst manufacturers, that are not fully commercialized. All manufacturers have many available. It is worthwhile to consider these sources since subsequent steps are accelerated.

After the design phase, exploratory catalysts are made for testing. Although small-scale preparations using colloidal laboratory procedures are used, careful records should be maintained. This not only protects patent claims but aids future reproduction and scale-up by manufacturers.

An important feature at this stage is optimization of critical properties. Design only specifies active components, supports, and promoters. It does not provide guidance on parameters such as concentration of the active component, method of preparation, and temperature of calcination. These factors, outlined in detail in Chapter 6, must be determined empirically. Careful planning using statistical strategy is necessary, since there are many significant variables.

3.6. CATALYST TESTING

Allied with design and preparation, catalyst testing is the exploratory screening of candidate catalysts. This phase does not yield either kinetics or process variables but merely ranks performance. Bench reactors used should be as simple and rapid as possible, for many samples are usually tested.[86] For ease in operation and interpretation, model compound reactions are helpful. Thus, for example, cumene dealkylation is a model for catalytic cracking, and thiophene hydrogenolysis for hydrodesulfurization. Care must be taken to ensure that the model system does indeed parallel process performance.

Design, preparation, and testing should be treated as a unit with continuous feedback, leading to optimized performance. Characterization is important since proper evaluation of catalyst parameters is necessary. These methods and procedures are discussed in Chapter 7.

3.7. PILOT UNIT TESTING

This stage is reached when one or more successful candidates emerge from the catalyst design. It involves testing under simulated process conditions, using pilot or semicommercial units. The principal objectives in using pilot units are the following:

1. To test commercial-like formulations under process conditions to ensure that design requirements are achieved. These include activity, selectivity, diffusional, and mechanical properties.

2. To optimize process variables, such as space velocity, temperature, and pressure. If activation (for example, reduction or sulfiding) is required, proper procedures are developed at this time.

3. To carry out long-term lifetime tests. The catalyst encounters components of the feed for the first time. Although the designer may have anticipated the effect of substances such as poisons, sensitivity of the catalyst formulation can only be checked with experiment. For well-behaved processes, lifetime tests of several hundred hours are necessary before long-term

trends become apparent. Longer periods are desirable but are seldom justified economically.

4. To develop scale-up data for commerical plant design. Economic evaluations based on projected plant configurations and unit construction are speculative up to this point. Pilot units studies provide more realistic data so that refinement of earlier estimates is possible. Also, any new reactor designs or innovations can be tested.

5. To obtain product samples. Large quantities of products may be accumulated for specification assessment. This is useful not only for testing effectiveness of the catalytic process but also in providing data for design of downstream components, safety, or market development, and so on.

6. To train personnel. Operators and engineers for the commercial units may be trained on pilot units, where the range of "adjustable" conditions is much wider and the result of misjudgments less disastrous.

Pilot units are investments and commitments of time and money. They are expensive and should not be misused. As much technical information as possible should be obtained during the laboratory work and a clear realization of the objectives of the pilot program formalized.[87] The size of the pilot unit is a very important factor in construction and operating costs. Keeping in mind that testing commerical formulations is the key task, the reactor unit must be sized so as to simulate the final design but yet be within reasonable limits. Sometimes simulating the performance of larger beds is necessary, e.g., to determine temperature profiles in adiabatic reactors. Heat-transfer-limited processes are best tested with commercially sized tubes since heat-transfer characteristics are difficult to estimate.

Every application has a different set of constraints. The degree of uncertainty and risk factors in the final design are important in determining the size of intermediate pilot units. It is useful at this point to simulate performance with computer models that may then be "calibrated" with pilot unit data. Such models provide very effective tools for parametric studies and assessment.[88]

The catalysts used in pilot units should be manufactured with commerical procedures. Large, consistent quantities of the catalyst are required. If the organization conducting the pilot program does not have access to manufacturing facilities, it is wise to engage the cooperation of catalyst manufacturers at this stage.

Automation of pilot units increases construction but reduces operating costs. Turnkey units are available for almost any size and application. However, before embarking on this route, care should be taken to ensure that the critical objectives of the program are achieved.

We hope, at this point, that pilot unit studies justify catalyst design. Wrong decisions in defining objectives will certainly be obvious. Unexpected problems always emerge at this stage. For example, a need for increased activity or selectivity may have been successfully treated, but the exposure to real feeds or process conditions indicates serious deactivation. A new cycle of design then evolves as this new threat is considered.

3.8. PROCESS DESIGN

The catalyst development is now ready for incorporation into the overall process design, but the story seldom stops there. After a period of long-term commerical usage, the catalyst may return with a different need encountered, and a new round of research, design, testing, and piloting begins. Loss of potassium in high-pressure steam reforming is an example. It was discovered that potassium, added to decrease carbon fouling in naphtha reforming, volatilizes too readily from the catalyst.[49] To counter this, catalyst designers incorporated an appropriate amount of an isoluble potassium silicate compound, which is only slowly hydrolyzed and so maintains effective levels of volatile alkali at all times. This innovative "time release capsule" was exactly the medicine required to extend catalyst life.

In the following chapters, the background to all of these steps is given in detail. The reader should at all times consider each area in perspective with the integrated development. Never forget the objective: to solve a process need with efficiency and economy in a reasonable time.

CATALYTIC MATERIALS
What They Are

It has been said that everything is a catalyst for something. Although profound, the statement is not very useful unless materials are organized into groups with common properties, explained with theories or models, and systematized into patterns from which new catalysts may be predicted. In this chapter we examine common types of catalytic materials, current theories underlining their mode of action, and activity patterns useful in design. Much of this is brief by necessity, but the interested reader will find sufficient references for further study. For the casual reader, this chapter illustrates the complex background in catalysis and testifies to the current attempts to lift catalysis from an "art" to a "science."

4.1. TYPES OF CATALYTIC MATERIALS

Catalytic materials fall into well-defined categories. Although we use broader classifications than those given in Chapter 1, the motivation is the same—to group according to common types of activity and to explain the catalytic behavior on the basis of common properties.

Table 4.1 is a list of catalytic materials with examples. The state of each catalyst is a consequence of process demands, e.g., for high activity, or degree of interaction with other components. Classification by electrical conductivity, as metals, semiconductors, and insulators, remains a satisfactory method in treating the theoretical background and behavior patterns of these widely differing materials.

TABLE 4.1. Types of Catalytic Materials

Type	State	Examples
1. Metals	Dispersed	Low: Pt/Al_2O_3, Ru/SiO_2
		High: Ni/Al_2O_3, $Co/kieselguhr$,
	Porous	Raney, Ni, Co, etc.
		$Fe-Al_2O_3-K_2O$
	Bulk	Pt, Ag gauze
2. Multimetallic		
clusters, alloys	Dispersed	$(Pt-Re, Ni-Cu, Pt-Au)/Al_2O_3$, etc.
3. Oxides	Single	Al_2O_3, Cr_2O_3, V_2O_5
	Dual, co-gels	$SiO_2-Al_2O_3$, $TiO_2-Al_2O_3$
	Complex	Ba TiO_3, $CuCr_2O_4$, Bi_2MoO_6
	Dispersed	NiO/Al_2O_3, MoO_3/Al_2O_3
	Cemented	$NiO-CaAl_2O_4$
4. Sulfides	Dispersed	MoS_2/Al_2O_3, WS_2/Al_2O_2
5. Acids	Dual, co-gels	$SiO_2-Al_2O_3$
	Crystalline	Zeolites
	Natural clays	Montmorillonite
	Promoted acids	Super acids SbF_5, HF
		Supported halides
6. Bases	Dispersed	CaO, MgO, K_2O, Na_2O
7. Other compounds	Chlorides	$TiCl_3-AlCl_3$
	Carbides	Ni_3C
	Nitrides	Fe_2N
	Borides	Ni_3B
	Silicides	TiSi
	Phosphides	NiP
8. Other forms	Molten salts	$ZnCl_2$, Na_2CO_3
	Anchored homogeneous catalysts	
	Anchored enzymes	

4.2. METALS

Over 70% of known catalytic reactions involve some form of metallic component.[44] Industrially, metals are used in catalytic reforming, hydro-cracking, ammonia and methanol synthesis, indirect coal liquefaction, oxidation, and a vast number of organic hydrogenation and dehydrogenation processes. Academically, metals are favored for research since they are easily prepared in pure form and conveniently characterized. In fact, most of the fundamental information leading to conceptual theories in catalysis originated with studies on metal systems.

The periodic table for transition metals is shown in Fig. 4.1. The periodic table is useful since catalytic behavior, like other chemical properties,

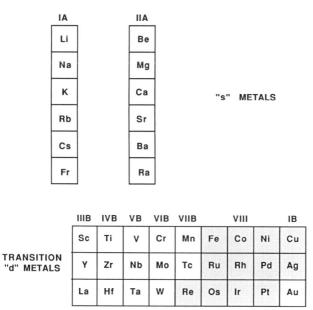

Figure 4.1. Transition metals in catalysis.

follows the same trends in going from right to left in each period and has similarities with other metals within the same group. It is not surprising that, when systematic catalytic results began to appear, all these properties were correlated against each other, some successfully, others not.[44] Naive as it may appear, this approach emphasizes the common origin of chemical behavior—electron configuration.

Successful catalytic applications are found only in d-electron transition metals. Alkali and alkaline s-metals revert too easily to ionic states under catalytic conditions and are found primarily as promoters. Some examples of catalysis with molten alkali metals exist, but they are not industrially relevant.[89]

Rare earth f metals likewise are too difficult to produce and too reactive to remain in the metallic state. Although widely found as oxide promoters and supports,[90] there is no example of successful catalysis with rare earth metals.

It is now known that d-electrons and their orbitals, hybrid and otherwise, are responsible for the bonding within the metal and at the surface. The type of bond in the bulk leads to properties such as crystal structure and dimensions, melting temperature, mechanical strength, magnetic state, and electrical conductivity. Surface bonds determine adsorption and surface mechanisms. The ability of a molecule to bond with the surface depends upon two factors[23]: (1) geometric or ensemble, and (2) electronic or ligand.

Much has been learned in recent years about the structures of metal surfaces,[91] which do not always parallel the crystallography of the bulk material.[92] Well-defined single-crystal surfaces provide us with an atomic view that is helpful in deciphering similar structures existing on dispersed metals.[93,94,95]

Ensemble effects are useful when adsorption requires a special grouping of surface atoms. To explain this, let us examine the simple example of ethylene adsorption on nickel, which occurs in a di-adsorbed mode.[96] Two nickel atoms, the right distance apart, are needed to bond a pair of carbon atoms. The bonds must be stable, but not too strong or subsequent reaction is difficult. Figure 4.2 shows symmetry and distances for low index planes of the face centered cubic nickel surface.

Two bond distances, 0.25 and 0.35 nm, are found. An ethylene molecule chemisorbed acrosss the 0.25 nm spacing results in Ni-C-C angles of 105°, close to the tetrahedral 109° and thus fairly stable and strong. On the other hand, the 0.35-nm distance results in an angle of 123°, which produces a

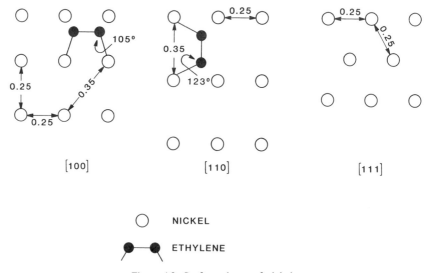

Figure 4.2. Surface planes of nickel.

weaker bond. Since this mode is not so strongly held and presumably is more active, planes with 0.35-nm ensembles, such as [100] and [110], are expected to show higher activity than those without, e.g., [111]. Also, atomic separation is a function of metal–metal distances in the unit cell, leading to the "volcano" curve, shown in Fig. 4.3.

Such a simple treatment illustrates the essential features of the geometric or multiplet theory.[96] This model was used to explain many types of reactions, now called demanding or structure sensitive,[97] that require unique groupings. Unfortunately, it was possible to explain any trend, and useful predictions did not result. Also, as more sophisticated surface structure determination became possible, it was clear that surface symmetry and dimensions do not always follow those of the bulk. The geometric theory, in its original form, is no longer fashionable. However, elements of this model survive today in the concept of reaction ensembles, believed to be decisive in explaining a large number of hydrocarbon reactions over metal films and high-area supported catalysts.[95] In addition to geometric groupings, this approach considers the symmetry of emerging orbitals that interact with adsorbing molecules. These features will be discussed after we first consider the historical developments in the competing electronic theory.

Based on the rigid band theory,[98] the electronic theory considered the metal as a collective source of electrons and electron holes, characterized

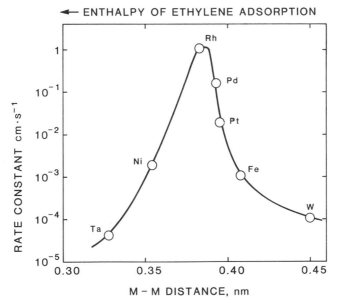

Figure 4.3. Ethylene hydrogenation as a function of atomic distance.[24]

by the band structure shown in Fig. 4.4. Moving from right to left across the periodic table means less d-electrons are available to fill the bands of energies corresponding to collectivized d-orbitals. Levels are filled successively until the Fermi level is reached. A certain number of vacant levels or d-holes are available for bonding with adsorbates; the lower the Fermi level the larger the number and stronger the adsorption. Thus, another version of the ethylene hydrogenation volcano curve, shown in Fig. 4.5, results. Ethylene hydrogenation rate is presumably related to the strength of adsorption through the number of d-holes.[71] This theory began to lose credibility when experiments on well-characterized Ni–Cu alloys become possible. The rigid band model predicts that alloying with Cu should result in some "intermediate" atom as nickel accepts electrons from copper. The d-band fills and the Fermi level increases until, at 55% Cu, no more holes are available. Catalytic activity, such as ethylene hydrogenation in Fig. 4.5, should be zero at this point. However, measurements on Ni–Cu alloys with well-characterized surface compositions,[100] shown in Fig. 4.6, do not support this.

Much of the earlier work on alloy films[24] was confused by the fact that binary systems are very susceptible to surface enrichment by one of the components.[101] Techniques for measuring alloy surface compositions have led to a much greater understanding.[102,103,104] Current views are that

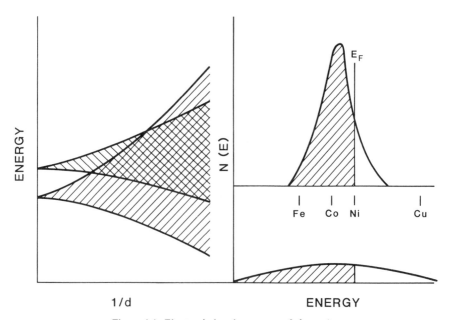

Figure 4.4. Electronic band structure of d-metals.

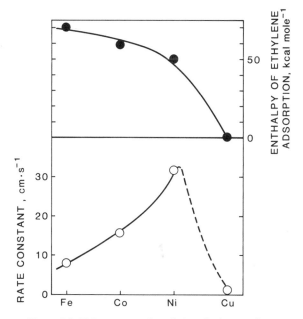

Figure 4.5. Volcano curve for ethylene hydrogenation.

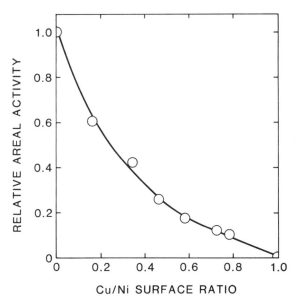

Figure 4.6. Cyclopropane hydrogenation/hydrogenolysis on Ni–Cu catalysts.[100]

atoms, such as copper, merely dilute the surface, so that the number of ensembles decreases in a manner consistent with probability statistics. Thus in Fig. 4.6, nickel retains its electronic character and activity drops as the number of adjacent nickel atoms decreases with copper addition.

Historically, as the geometric and electronic theories began to fade, a new model emerged, incorporating features of both. A localized bond theory, the model is based on concepts in the ligand field treatment of bonding in inorganic and organometallic complexes. Face-centered cubic nickel has octahedral symmetry with its neighbors, whose electric field splits the fivefold degenerate d-orbitals into two groups.[105] The higher energy group (e_g) consists of $d_{x^2-y^2}$ and d_{z^2} orbitals, the lower (t_{2g}) d_{xy}, d_{xz}, and d_{yz}. These are shown in Fig. 4.7. Centered at atomic positions in the same way as in the geometric theory, these orbitals emerge at different angles from the surface plane, as shown in Table 4.2.

Figure 4.8 illustrates how these differences influence available orbitals.[106] In the [100] plane, e_g orbitals emerge perpendicular and parallel to the surface, t_{2g} orbitals at 45°. Empty or partially filled e_g orbitals overlap the $1s$ orbital of hydrogen in two locations, with one e_g orbital at the "on top" position or with five at the position one-half of an atomic layer into the surface. Similar situations prevail for other surface planes. With this model, it is now possible to visualize interactions of molecules such as hydrogen and ethylene with the nickel surface. Figure 4.9 shows the possibilities. Currently these models are only qualitative and much more

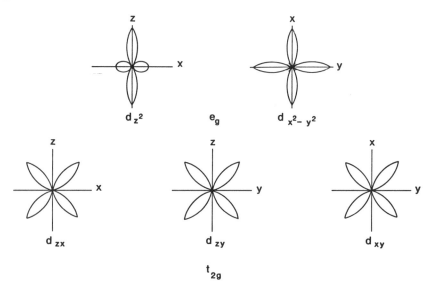

Figure 4.7. Crystal field splitting of d-orbitals.

TABLE 4.2. Orbital Angles on Surface Planes

		Plane		
	Orbital	[100]	[110]	[111]
eg	d_{z^2}	90°	45°	36°16'
eg	$d_{x^2-y^2}$	0°	45°	36°16' (2)
t_{2g}	d_{xy}	0°	30° (2)	30°
t_{2g}	d_{xz}	45° (2)	30° (2)	30°
t_{2g}	d_{yz}	45° (2)	90°	30°

research is necessary to assign quantitative values for heats of adsorption of these modes. However, the theory in principle explains (1) different modes and heats of adsorption, (2) differences in reactivities between planes, (3) patterns in different metals, and (4) alloy effects.

It is known that heat of adsorption decreases with coverage.[24] Although there are many explanations, site heterogeneity is most plausible. Adsorption occurs first on the strongest sites until they saturate, followed by bonding with the weaker sites. The fact that reaction may occur only on selective locations complicates the interpretation of areal rates deduced from measurement of full-coverage chemisorption.

Careful measurements of reactivities on planes have revealed differences. For examples, Table 4.3 shows such a dependence in ammonia synthesis.[107]

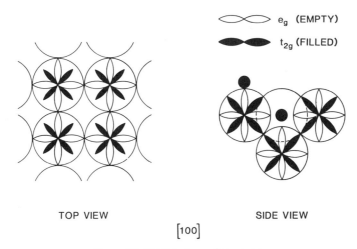

e_g (EMPTY)

t_{2g} (FILLED)

TOP VIEW   SIDE VIEW

$$[100]$$

Figure 4.8. Orbitals for surface atoms.

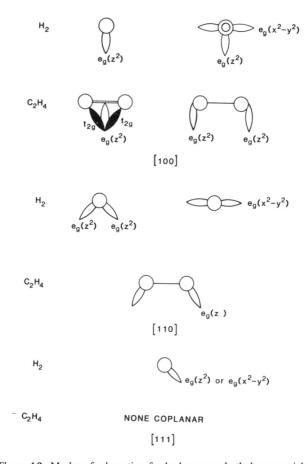

Figure 4.9. Modes of adsorption for hydrogen and ethylene on nickel.

TABLE 4.3. Differences
in Crystal Planes
Ammonia Synthesis [107]

Plane	Activity
[100]	21
[110]	1
[111]	440

The model is consistent with observed patterns in adsorption and reactivity across the periodic table. Both spacing and orbital occupancy are involved, so that combinations of ensemble and ligand effects are necessary to explain dependencies.

Finally, alloying patterns are justified. Figure 4.6 shows cyclopropane hydrogenation decreasing uniformly with surface Cu concentration. Each nickel atom retains its orbital integrity, and the effect of copper is simply to dilute. For instance, if n surface nickel sites are required to generate an absorbed species, then a surface copper fraction f_{Cu} results in an activity decrease, A/A_0, given by

$$\frac{A}{A_0} = (1 - f_{Cu})^n \qquad (4.1)$$

An additional feature of the localized bond model emerges. Small crystallites of metals exist as pseudospherical cubo octahedra[108] as shown in Fig. 4.10.

In the ideal form of this structure, only [100] and [111] planes occur, but atoms are also situated at sites corresponding to corners and at edges between these planes. These have lower coordination than plane sites and exhibit different orbital symmetries. The distribution of these sites is shown in Fig. 4.11, suggesting that adsorption and reaction involving low coordination sites increases as crystallite size decreases, whereas those occurring on face sites increase with size.[81] Thus a crystallite size effect is seen, as verified by experiments on selectivity factors in cyclohexane reactions

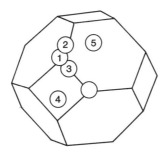

1	CORNER SITES
2	EDGE SITES, $[\bar{1}11],[1\bar{1}1]$
3	EDGE SITES, $[\bar{1}11],[\bar{1}00]$
4	FACE SITES, $[\bar{1}00]$
5	FACE SITES, $[\bar{1}11]$

Figure 4.10. Surface sites on cubo-octahedra.

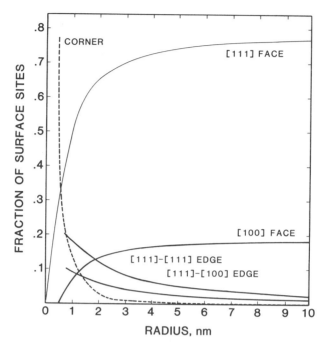

Figure 4.11. Distribution of surface sites with crystallite size.[108]

(Fig. 4.12).[110] Real crystallites are not ideal cubo-octahedra. Imperfect planes develop other types of sites, as shown in Fig. 4.13.[83] Low coordination sites such as corners, edges, steps, and kinks are high energy sites. There is evidence that reactions such as hydrogenolysis occur preferentially on these sites,[95] so that selectivity patterns may depend on crystallite size or surface morphology. Also, blocking promoters, such as Cu in Ni, and poisons, such as S on Pt, bond with these sites and account for the pronounced effect of these agents in decreasing hydrogenolysis.[83] Experiments such as these have led to much interest in the theoretical properties of these small crystallites.[111,112]

The question of strong metal support interactions (SMSI) has been mentioned in Chapter 2. There is little question that the effect exists, but the mechanism is still far from clear.[63] For metals supported on easily reducible oxides such as TiO_2, reduction at high temperatures results in loss of hydrogen adsorption and reactivity. Mild oxidation reverses the loss, verifying speculation that Ti^{3+} sites are generated and, in some way, poison orbital participation. Similar but less pronounced effects are observed with common supports such as Al_2O_3 and SiO_2. Many examples of SMSI effects and possible explanations are given in reviews.[59,63] Although interesting

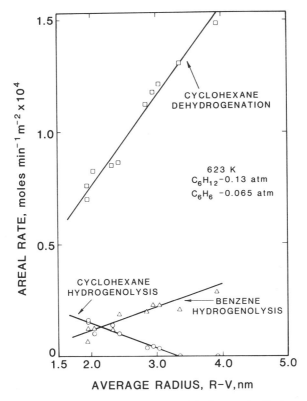

Figure 4.12. Crystallite size effects of cyclohexane dehydrogenation/hydrogenolysis.[110]

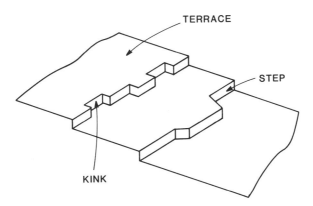

Figure 4.13. Steps and kinks in crystal planes.[83]

TABLE 4.4. Experimental Catalytic Activity of Metals[a]

Group	Metal	Reaction
1B	Cu	Water gas shift, CH_3OH synthesis
		Hydrogenation of aromatic nitro compounds to aromatic amines
		Hydrogenation of aldehydes and ketones to alcohols
		Hydrogenation of atomic olefinic groups
		Hydrogenation of unsaturated nitrites to unsaturated amines
	Ag	Hydrogenation of diolefins and alkynes to monoolefins
		Hydration of ethylene oxide alkylation of aromatics with olefins oxidation of alcohols to aldehydes
		Oxidation of ethylene to ethylene oxide
		Synthesis of HCN from NH_3 and CH_4
		Oxidation of CH_3OH to HCHO
	Au	
VIII	Ni	Hydrogenation of olefins, aromatics, anilines, phenols, nitriles, naphthalenes, alcohols, and aldehydes, CO, and CO_2 (Fischer–Tropsch)
		Dehydrogenation of cyclohexanols, cyclohexanones, alkanes, alcohols
		Hydrogenolysis of C–C, C–N, N–O, steam reforming of CH_4
	Pd	Hydrogenation of olefins,
		aromatic aldehydes and ketones,
		aliphatic alcohols, aldehydes, ketones,
		unsaturated nitriles,
		aromatic nitro compounds to aromatic amines,
		phenols
		Dehydrogenation of cyclohexanes, cyclohexenes
		Oxidation of hydrocarbons
		CH_3OH synthesis
	Pt	Hydrogenation of olefins, dienes, acetylenes,
		aromatics aliphatic aldehydes
		nitro aromatics,
		naphthalenes
		Dehydrogenation of cyclohexanes, cyclohexenes,
		cyclohexanols, cyclohexanones, alkanes
		Deformylation of aldehydes
		Oxidation of hydrocarbons, CO, NH_3
		Reduction of nitrogen oxides
VIII_2	Co	Hydrogenation of olefins, aldehydes, nitriles, atomic amines, CO, CO_2 (Fischer–Tropsch)
		Hydrogenolysis of C–C
		Isomerization of olefins
	Rh	Hydrogenation of olefins, aromatics, nitriles, phenols, ketones, nitrobenzenes, CO_2 reforming of CH_4
	Ir	
VIII_1	Fe	NH_3 synthesis
		Hydrogenation of CO, CO_2 (Fischer–Tropsch)
		Dehydrogenation of cyclohexanes
		Conversion of alcohols and NH_3

(*continued*)

TABLE 4.4. (*continued*)

Group	Metal	Reaction
	Ru	NH$_3$ synthesis
		Hydrogenation of CO, CO$_2$ (Fischer–Tropsch), olefins, oximes, nitriles, phenols, benzyl alcohols, aryl amines, aromatic heterocycles, cyclopentane, aliphatic ketones, nitrobenzenes, oxidation of hydrocarbons
	Os	Ammonia synthesis
VIIB	Mn	
	Tc	
	Re	Hydrogenation of olefins, aromatic, carboxylic acids, amides, nitrogenzenes, ketones, dehydrogenation of cyclohexanes, alcohols

^{a} Note: Reverse reactions are also catalyzed. Similarities within groups are expected.

from a fundamental point of view, the role of these phenomena in practical catalysis remains to be demonstrated.

Another emerging area of investigation is the behavior of supported metal cluster compounds.[113,114] For research purposes, these materials provide well-defined models. However, they have not yet led to any major developments in industrial applications.

We leave the topic the metal catalysis with a listing of observed activities (Table 4.4). Patterns of behavior are also important. Examples of these are given in Chapter 5, and the reader should also refer to many excellent sources in the literature.[115]

4.3. SEMICONDUCTORS

Catalytically interesting semiconductors are oxides and sulfides of transition elements.[45,116,117] Unlike metals, electronic bands do not overlap but form separated regions, shown in Fig. 4.14.

The lower band, the valence band, contains levels normally filled by electrons participating in valence bonds of the crystal.[99] Separated by an energy E_g, the second band is empty unless electrons are promoted by heat or radiation from their valence positions. The upper band is called the conduction band, since electrons with these energies are sufficiently free to migrate with the application of an electric field, i.e., conduction. Simultaneously, the "hole" left by the promoted electron conducts electrical energy in the opposite direction. Two types of current carriers are involved. Semiconductors of this type are called intrinsic but are not important in catalysis, since the temperature range, 300–700°C, is usually too low to generate a sufficient number of promoted electrons.

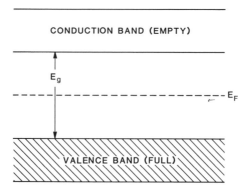

Figure 4.14. Electronic structure of an intrinsic semiconductor.

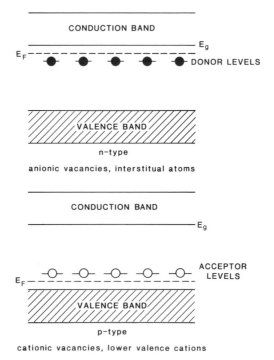

Figure 4.15. Electronic structure of impurity semiconductors.

Much more significant are the impurity n-type and p-type shown in Fig. 4.15. In an n-type semiconductor, for example ZnO, the structure is nonstoichiometric, and an excess of Zn^+ ions exists. The small cation is easily accommodated in interstitial positions, generating energy levels that are filled (donor) with energy values close to the conduction band. Only a small amount of thermal or radiation energy is necessary for promotion and conduction. Conductivity occurs via electrons in the conduction band.

With p-type semiconductors, there is an excess of anions and cationic vacancies. These produce empty levels (acceptor) close to the valence band, from which an electron is easily promoted. Conductivity occurs through the resulting hole in the valence band.

Donor and acceptor levels are also created by introducing (doping) altervalent cations into interstital or cationic lattice positions. Both may coexist within the same crystal.

The Fermi level is the electrochemical potential, midway between the highest filled and the lowest empty levels. Since the Fermi level also responds to doping (increases with more donor levels, decreases with acceptor levels) and is easily measured, it is a convenient index with which to follow changes in electronic structure.

Table 4.5 gives the type of semiconductivity found in common transition oxides and sulfides. Many of these will be recognized as catalytically important materials.

The role of these electronic structures in catalytic reactions is best demonstrated with a much-studied probe reaction, decomposition of nitrous oxide[118]:

$$2N_2O \rightarrow 2N_2 + O_2 \tag{4.2}$$

The mechanism is believed to proceed as follows:

$$N_2O + e \rightarrow N_2 + O_a^- \tag{4.3}$$

$$O_a^- + N_2O \rightarrow N_2 + O_2 + e \tag{4.4}$$

where donation of electrons to the catalyst is the rate-determining step. Transfer of an electron from adsorbed O^- only takes place if the Fermi level of the surface is below the ionization potential of O^-. This is most likely to occur in p-type semiconductors, where the Fermi level is sufficiently low as shown in Fig. 4.16.

Relative activities for reaction (4.2) over a wide range of semiconductors are given in Table 4.6. Several important points emerge from the data in Table 4.6. First, p-type oxides are more active than n-type oxides. This is consistent with the model of electron transfer to the catalyst. Second, the

TABLE 4.5. Types of Transition Element Semiconductors

Group	Element	Compound	Type	$E_g e_v$
1B	Cu	CuO	n, p	1.4
		Cu_2O	p	1.9
		Cu_2S	p	1.7
	Ag	AgO	p	0.7
		Ag_2O	p	1.5
		Ag_2S	n	1.2
	Au	Au_2O_3	n	—
VIII_3	Ni	NiO	p	2.0
		Ni_2O_3	n	—
		NiS	p	—
	Pd	—		
	Pt	PtO_2	n	—
VIII_2	Co	CoO	p	0.8
		Co_2O_3	n	—
		Co_3O_4	p	0.9
	Rh	—		
	Ir	—		
VIII_1	Fe	FeO	p	0.4
		$\gamma\text{-}Fe_2O_3$	n	1.0
		$\alpha\text{-}Fe_2O_3$	n	2.2
		Fe_3O_4	n	0.4
		FeS	p	0.1
		FeS_2	p, n	1.2
	Ru	RuO_2	—	0.1
	Os	—		
VIIB	Mn	MnO	p	1.25
		$\gamma\text{-}MnO_2$	n	0.3
		$\beta\text{-}MnO_2$	n	0.6
		—		
	Re	ReO_2	—	1.2
VIB	Cr	CrO	p	—
		CrO_2	n	0.3
		CrO_3	n	3.2
		Cr_2O_3	p, n	1.9
		CrS	n	0.9
	Mo	MoO_3	n	2.9
		MoS_2	p	1.2
	W	WO_2	n	5.0
		WO_3	n	2.8
		WS_2	p	—
VB	V	V_2O_3	—	0.3
		V_2O_5	n	2.1

(*continued*)

TABLE 4.5. (*continued*)

Group	Element	Compound	Type	$E_g\, e_v$
	Nb	NbO	p	
		Nb_2O_5	n	3.3
	Ta	Ta_2O_8	n	3.6
IVB	Ti	TiO	p	—
		TiO_2	n	3.1
		Ti_2O_3	p	2.2
	Zr	ZrO_2	n	2.3
	Hf	HfO_2	n, p	4.4
IIIB	Sc	Sc_2O_3	p, n	2.2
	Y	Y_2O_3	n	2.9
	La	La_2O_3	n	2.6
	Al	$\gamma\text{-}Al_2O_3$	n	7.3
	Zn	ZnO	n	3.3
		ZnS	n	3.6
	Mg	MgO	n, p	8.7
	Ca	CaO	n, p	7.5
	Ce	CeO_2	n	—
	Th	ThO_2	n	3.5

overall trend does not follow any specific quantitative correlation. Certainly, other factors such as dispersion and impurities must influence the results, but dimensions and symmetry at adsorptive sites are also important. Refinement beyond mere type of semiconductivity is necessary.

Third, for a given crystal structure, the position of the Fermi level plays an important role. For example, in three NiO-containing samples in Table 4.6, the influence of doping is clearly evident. Addition of Li^+ ions to the Ni^{2+} lattice creates more acceptor sites, decreasing the Fermi level and increasing activity. With Cr^{3+} incorporation, donor levels are produced, the Fermi level increases, and activity decreases.

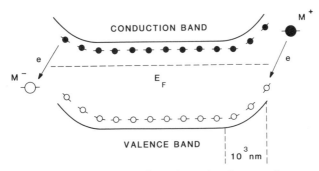

Figure 4.16. Electron transfer at the semiconductor surface.

TABLE 4.6. Relative Activities for Nitrous
Oxide Decomposition

Oxide	Type	Eg	Relative activity
Cu_2O	p	1.9	10.8
CoO	p	0.8	7.91
NiO + 2% Li_2O			3.78
NiO	p	2.0	1.00
NiO + 2% Cr_2O_3			3.02×10^{-2}
CuO	n	1.4	7.28×10^{-1}
MgO	n	8.7	2.10×10^{-1}
CaO	n	7.5	1.10×10^{-1}
CeO_2	n	—	7.10×10^{-2}
Al_2O_3	n	7.3	2.75×10^{-2}
ZnO	n	3.3	1.25×10^{-2}
TiO_2	n	3.1	9.48×10^{-3}
Cr_2O_3	n	1.9	7.30×10^{-3}
Fe_2O_3	n	2.2	5.30×10^{-3}

The most successful treatment of semiconductor catalysis was given by Vol'kenstein, and the serious reader is referred to his publications.[118,119,120] Although phenomenological and qualitative in nature, Vol'kenstein's model explains observed trends and predicts important results. For example, in C_2H_5OH decomposition, dehydrogenation is predicted to increase with increasing Fermi level, whereas dehydration decreases. Table 4.7 shows that this is true qualitatively, with dehydrogenation increasing and dehydration decreasing as the oxide becomes more n-type.

Another prediction of Vol'kenstein's theory is that crystallite size influences semiconductor electronics. Figure 4.16 shows the bending of energy levels at the surface, the so-called Debye length. Since catalytic crystallites are much smaller than this dimension, the phenomenon is not found in these materials.[119] Catalysts may be treated as two-dimensional semiconductors, a fact that greatly simplifies theoretical models.

Still another consequence of the Vol'kenstein theory is photocatalysis.[121,122] It is well known that light quanta excite electrons out of and into impurity levels. The effect is to change the Fermi level, with predicted influences on catalytic reactions.[123,124] In recent years, a large amount of attention has been directed toward photocatalysis with semiconductors as a means to splitting water into hydrogen and oxygen.[125]

From a practical viewpoint, the most significant features of the electronic model for semiconducting catalysis is the framework around which correlations may be organized. One such pattern is given in Table 4.7, others

TABLE 4.7. C_2H_5OH Selectivity[a]

Oxide	Percent C_2H_5OH decomposed to	
	$H_2 + C_2H_4O$	$H_2O + C_2H_4$
γ-Al_2O_3	1.5	98.5
Cr_2O_3	9	91
TiO_2	37	63
ZrO_2	55	45
Fe_2O_3	86	14
ZnO	95	5

[a] Reference 121.

in Chapter 5. Other correlations of this type are given in the literature.[45] Data are always difficult to find and many studies were made under different conditions, in which dispersions were not always measured. Nevertheless, in cataloging activity patterns for specific reactions, much more validity is possible within the guidelines of this electronic treatment.

For fundamental understanding of catalytic sites, this crude treatment is not enough. Some consideration of surface geometry and orbital availability, in much the same way as in metal catalysis, will be necessary for greater understanding.[126] This is one area of scientific catalysis awaiting exploitation.

Reactions catalyzed by oxides and sulfides are listed in Table 4.8.

4.4. INSULATORS AND SOLID ACIDS

Insulators are semiconductors with large values of E_g and very low concentrations of impurity levels. Electrons remain localized in valence bonds so that redox-type reactions, found with better conducting metals and oxides, do not occur. However, these materials exhibit sites that generate protons, thereby promoting important carbonium ion reactions such as cracking, isomerization, and polymerization. Acid solids are characterized as (1) single oxides, such as Al_2O_3 and SiO_2, (2) natural clays, (3) mixed amorphous oxides, for example SiO_2–Al_2O_3, and (4) zeolites. Table 4.9 shows the wide range of acid strengths encountered by these and other oxides.

4.4.1. Alumina

The origin of acid sites is best illustrated with γ-Al_2O_3, which is often used as a support but has inherent uses of its own.[127] Prepared as a hydrous

TABLE 4.8. Catalytic Activity of Transition Metal Oxides and Sulfides

Group	Element	Compound	Reaction
1B	Cu	CuO	Oxidation of Co, hydrocarbons nitric oxide
		$CuCr_2O_4$	Hydrogenation of nitro aromatic aromatic carbonyls, aromatic olefins, unsaturated nitriles Decarboxylation of acids Synthesis of pyrazines Decomposition of perchloric acid
		AgO	—
		Au_2O_3	—
VIII_3	Ni	NiO	Dehydrogenation of alkanes Oxidation of hydrocarbons
VIII_2	Co	CoO	Hydrogenation of olefins
VIII_1	Fe	Fe_3O_4	Water gas shift Dehydrogenation of cyclohexanes Conversion of alcohols to amines
VIIB	Mn	MnO_2	Oxidation of alcohols Dehydrogenation of ketones Hydration of nitrites
VIB	Cr	Cr_2O_3	Cyclization of C_3 to C_5 paraffins
	Mo	MoO_3	Dehydrogenation of cyclohexanes Dehydration of aliphatic alcohols Oxidation of propylene Polymerization of olefins Disproportionation of olefins Oxidation of CH_3OH to CH_2O
		MoS_2	Water gas shift Hydrogenolysis of C–S, C–N, C–O
	W	WO_3	Disproportionation of olefins Dehydration of alcohols Cyclization of C_3 to C_5 paraffins
		WS_2	Hydrogenolysis of C–S, C–N, C–O
VB	V	V_2O_5	Oxidation of hydrocarbons Butene to maleic anhydride Benzene to pthalic anhydride Cyclization of C_3 to C_5 paraffins
IVB	Ti	TiO_2	Photocatalytic decomposition of water

TABLE 4.9. Acidity of Catalytic
Materials

Solid	pK_a range
$SiO-Al_2O_3$	< -8.2
Montmorillonite clay	-5.6 to -8.2
Kaolinite clay	-5.6 to -8.2
γ-Al_2O_3	$+3.3$ to -5.6
$SiO-MgO$	$+3.5$ to -2.5
SiO_2	-2.0
TiO_2	$+6.8$ to $+1.5$
$MgAl_2O_4$	>7.0
Cao	>7.0
MgO	>7.0

oxide, γ-Al_2O_3 is activated by calcination above 300°C.[64] The following
sequence occurs:

$$\begin{array}{ccc} OH & OH & O^- \\ | & | & | \\ O-Al-O-Al & \xrightarrow{-H_2O} & -O-Al^+-O-Al \\ & & \text{Lewis} \quad \text{Basic} \\ & & \text{site} \quad \text{site} \end{array}$$

$$\underset{-H_2O}{\overset{+H_2O}{\rightleftharpoons}} \quad \begin{array}{cc} OH & O^- \\ | & H^+ | \\ -O-Al-O-Al \\ \text{Bronsted} \\ \text{site} \end{array} \qquad (4.5)$$

Dehydration of the hydrous oxide gives a surface containing both Lewis
(electron acceptor) and basic sites. However, sufficient H_2O is always present
to generate Bronsted sites. The proton is hydrogen-bonded to the hydroxyl
group with a strength related to the electronic environment of the Al^{3+} ion.
Factors that tend to shift the electron distribution in the Al-OH bond toward
the Al^{3+} ion weaken the OH-H^+ bond, making the proton more accessible,
and increases the acidity. Aluminum ions in Al_2O_3 occupy both tetrahedral
and octahedral sites with neighboring oxygens. Surface coordination
depends on crystal faces exposed, so that a number of schemes are possible.
One representation is shown in Fig. 4.17.[128] Numbers 1-5 identify different
types of isolated hydroxyl ions and their relationship to the oxygen and
Al^{3+} ions in the layer below the surface. Clearly, site 1 has a weaker OH-H^+
bond and greater acidity than other sites. A distribution of acidities then
develops, itself an important factor in characterizing acidic materials.

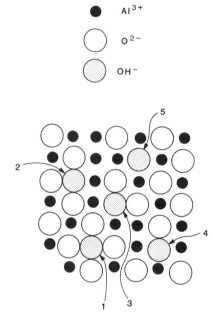

Figure 4.17. Suggested scheme for acidic sites on γ-Al$_2$O$_3$.[127]

Neutralization of the proton, H$^+$, by a base, B, occurs through the reaction[55]

$$B + H^+ \leftrightarrows BH^+ \qquad (4.6)$$

in which the acid ionization constant is

$$K_a = \frac{[H^+]\gamma_H[B]\gamma_B}{[BH^+]\gamma_{BH^+}} \qquad (4.7)$$

with γ's the appropriate activity coefficients. It is convenient to express acid strength by

$$pK_a = -\log_{10} K_a \qquad (4.8)$$

so that strong acidity shows large negative values of pK_a. Methods for measuring pK_a are discussed in Chapter 7. Distributions in γ-Al$_2$O$_3$ are given in Fig. 4.18, which also shows how acidity increases with calcination temperature as the surface becomes structured. Further heating above 400°C causes dehydration and a decrease in acidity.[55]

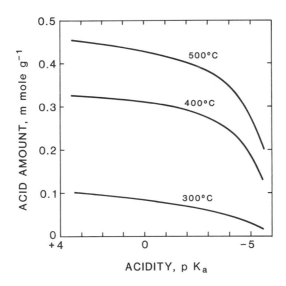

Figure 4.18. Acid strength distribution for γ-Al_2O_3.[55]

4.4.2. Natural Clays

Highly acidic natural clays, montmorillonite are complex layers of SiO_4 and AlO_4 tetrahedra. They also contain small amounts of MgO and Fe_2O_3. These impurities are leached with sulfuric aid, which also adds protons to increase maximum pK_a values from -3.0 to -8.2. These clays were the first cracking catalysts used with fixed and moving beds. However, they were quickly replaced by the superior synthetic silica–aluminas that were ideal for fluidized beds.[129,130] Today, they are used as the matrix in zeolite-based cracking catalyst.

4.4.3. Silica–Alumina

Clearly, single oxides such as Al_2O_3 and SiO_2 are much less acidic than combinations of the two. In fact, most highly acidic materials are luminosilicates. The reason for this is seen in the structure

$$
\begin{array}{ccccc}
 & H & & H & \\
 & | & H^+ & | & H^+ \\
O & O & O & O & O \\
 & \diagdown | \diagup & & \diagdown | \diagup & \\
 & Al^{3+} & & Si^{4+} &
\end{array}
\tag{4.9}
$$

in which Si^{4+} replaces tetrahedral Al^{3+} with a more electropositive center, thereby weakening the O-H bond and increasing acidity. Silicon and aluminum ions must be intimately mixed during preparation. This effect is shown in Fig. 4.19, where we see that optimum acidity develops at about 30% Al_2O_3. Synthetic silica–alumina catalysts led to fluidized catalytic cracking, the largest and most important process in the refining industry. Other dual oxide systems display degrees of acid behavior, but the superiority of silica–alumina justifies its role in catalytic cracking.[131]

Carbonium ion-type reactions initiated by these acidic oxides are shown in Table 4.10.

Activity patterns follow acid strength, as shown in Fig. 4.20. Parallel trends are obeyed for gasoline production and cumene dealkylation, and the latter is commonly used as a "model" reaction for catalytic cracking.

4.4.4. Zeolites

Zeolites, like clays and synthetic $SiO–Al_2O_3$ catalysts, are aluminosilicates. Unlike these materials, they have three properties that make them unique and deserving of a separate category. First, they are highly crystalline with well-defined structures.[132] The aluminosilicate framework encloses cavities occupied by large ions and water molecules. Access to these cavities of various sizes is through a network of openings ranging from 0.3 to 1.0 nm in diameter, which are of the order of molecular dimensions. Size and shape of these pores determine which molecules enter the cavities and which are

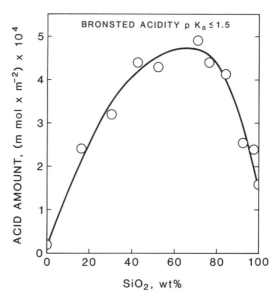

Figure 4.19. Acidity in silica–alumina catalysts.[131]

TABLE 4.10. Reactions
Catalyzed by Solid Acids

Cracking of C-C bonds
Alkyl group isomerization
Double bond isomerization
Dealkylation
Dehydroxylation
Dehydrogenation
Polymerization
Oligomization
Cyclization
Alkylation
"Coke" formation

excluded. Also called "molecular sieves," zeolites possess the ability to be shape and size selective in catalytic molecular rearrangements.[133]

Second, ions within the cavities are easily exchanged with a large number of altervalent ions. These ions exert large electrostatic or polarizing forces across the small dimensions of the cavity. Electron distribution shifts in acidic hydroxyl groups may be tailored to give acidities 10^4 times larger than in SiO_2-Al_2O_3. They are indeed "super acids."[134]

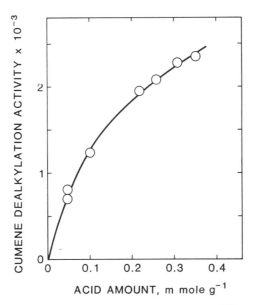

Figure 4.20. Activity patterns for acid solids.[129]

SOLID
TETRAHEDRON

TRUNCATED
OCTAHEDRON

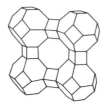

FACE OF CUBIC ARRAY
OF TRUNCATED OCTAHEDRA

Figure 4.21. Zeolite building blocks.

Third, ions introduced into cavities through ion exchange have separate activities of their own. Opportunities for dual functional catalysis involving acidity with other activities are greatly facilitated.[57] These three factors have made zeolites a powerful addition to the arsenal of catalytic materials.[135,136]

Zeolites are built from basic units consisting of four oxygen ions packed like spheres (Fig. 4.21). Interstices between the oxygen ions are filled with Al^{3+} and Si^{4+} ions in the ratio $x:y$. By joining these tetrahedra together through apical oxygen atoms, a truncated octahedron "sodalite" unit is produced. With this as a building block, a vast number of structures can be produced. For example, by joining them through the four-membered faces in a cubic array, the structure in Fig. 4.21 results. When two of these

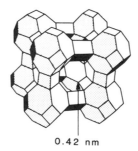

0.42 nm

Figure 4.22. Structure of type A zeolite.

are combined, as in Fig. 4.22, a cavity with 0.7-nm openings is produced. This is the zeolite type A structure, designed mostly for separation processes in which molecules less than 0.4–0.5 nm in size are preferentially adsorbed.

On the other hand, if the sodalite units are bridged across six-membered faces, the resulting structure (Fig. 4.23) has a 0.73-nm opening to a 12-nm cavity, forming faujasite. Still another arrangement gives the parallel pores of mordenite (Fig. 4.23).

There are over 70 known different kinds of building arrangements, each resulting in a distinct structure. Some of these occur as natural minerals; some are synthetic. The search for new combinations goes on. Catalytically important zeolites are listed in Table 4.11. More details on exact structures will be found in the references.[137,138,139] In recent years, much attention has been directed toward a new zeolite, ZSM-5, whose structure is shown in Fig. 4.24. This is a zeolite with openings of 0.54–0.56 nm, intermediate between the zerolite A (0.4 nm) and faujasite X and Y (0.74 nm). It has been successfully applied to the production of gasoline from methanol and for other specialized purposes.[140]

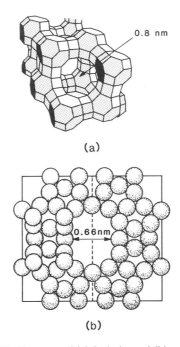

0.8 nm

(a)

0.66nm

(b)

Figure 4.23. Structure of (a) faujasite and (b) mordenite.

TABLE 4.11. Catalytic Zeolites

Zeolite	Pore opening	
	No. of O^{2-} ions	Dimensions (nm)
Faujasite, X, Y	12	0.74
Mordenite	12	0.74
Offretite	12	0.64
ZSM-5	10	0.54×0.56
Zeolite A	8	0.41
Erionite	8	0.36×0.52

These materials have the general formula

$$\text{Me}^{n+}_{x/n}[(\text{AlO}_2)_x(\text{SiO}_2)_y]m\text{H}_2\text{O} \tag{4.10}$$

where Me^{n+} is the cation that satisfies electric neutrality in the structure. During catalysis, the $m\text{H}_2\text{O}$ molecules are removed, leaving a large cavity in which the reaction takes place. Cationic positions vary from one structure to another, but some of them are found at locations within the internal surface of the cavities. As such, they play a vital role in establishing catalytic properties. Zeolites are most easily prepared from sodium silicates and aluminates, so that sodium is the cation in freshly made material. This is readily ion-exchanged with a large number of ions and the effect is very dramatic, as demonstrated in Table 4.12. There are various models to explain these results. In addition to different valences, the Lennard–Jones ionic radius decreases from Na^+ to H^+. This suggests that a good correlation exists with the factor ne/r, the polarizing power. This polarizing effect could be exerted directly across the cavity or indirectly through the zeolite lattice to neighboring OH groups.[141]

Shape and size selectivity are important when molecules approach the critical dimensions of the pore opening. In Table 4.13, the diameters of common hydrocarbons are listed in order of size, and we expect to find the largest effect in zeolites with pores and cages close to these dimensions.

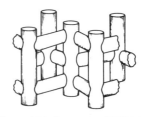

Figure 4.24. Structure of ZSM-5.

TABLE 4.12. Effect of Faujasite
Cation on Activity for
Cumene Dealkylation[a]

Cation	Relative activity
Na^+	1.0
Ba^{2+}	2.5
Sr^{2+}	2.0×10
Ca^{2+}	5.0×10
Mg^{2+}	1.0×10^2
Ni^{2+}	1.1×10^3
La^3	9.0×10^3
H^+	8.5×10^3
$SiO_2-Al_2O_3 = 1.0$	

[a] Reference 141.

There are three possible types of shape and size selectivity effects, as shown in Fig. 4.25. First, the reactant molecules may be too large to enter the cavities. Comparison of Tables 4.11 and 4.13 shows that all of the molecules access faujasite structures. Only molecules larger than pentamethyl benzene are excluded. Early examples of shape and size selectivity were almost completely limited to small openings and normal versus branched paraffins. For example, n-hexane was selectively cracked in the presence of 3-methylpentane over zeolite A catalysts. Other, more subtle, effects may occur. The diffusivity of trans-butane-2 is 200 times larger than that of cis-butene-2 in zeolite CaA. By adding Pt to CaA, selective hydrogenation of trans-butene-2 occurs.[134]

TABLE 4.13. Critical Diameter of Hydrocarbons

Hydrocarbons	Size (nm)
n-Paraffins	0.45
Methyl-paraffins	0.57
Dimethyl paraffins Benzene Toluene p-Xylene	0.63
o,m-Xylene 1,2,4-Tri-methyl benzenes 1,2,4,5-Tetra-methyl benzene naphthalene	0.69
1,3,5-Tri-methyl benzene Penta-methyl benzene	0.78

REACTANT
SELECTIVITY

PRODUCT
SELECTIVITY

RESTRICTED TRANSITION
STATE SELECTIVITY

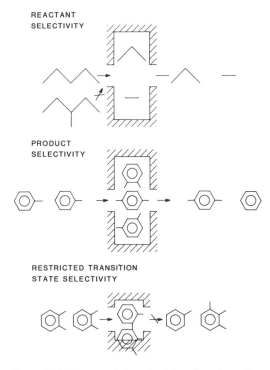

Figure 4.25. Shape and size selectivity effects in zeolites.

Development of the ten-ring ZSM-5 zeolites extended these promoters to more sophisticated selectivities.[140] Molecules of varying degrees of branching and chain length and alkyl aromatics now came within scope. Figure 4.26 shows the relative cracking rates of heptane isomers over ZSM-5, clearly demonstrating the effect of size exclusion. Since critical dimensions of alkyl benzenes are close to the pore opening, even small changes effect diffusivity, as demonstrated in Table 4.14.

Differences such as these are important in the second type of shape or size selectivity control, product exclusion. This is the case where undesired products cannot egress cavities, whereas the desired ones can. Toluene disproportionation to benzene and *p*-xylene is an example of successful applications of this principle.

A third type of control, called spatiospecificity, occurs when both reactants and products pass the opening but reaction intermediates or transition states are restricted by the size of the cavity. In xylene isomerization processes, selectivity is lost through disproportionation to toluene and trimethylbenzene. Diphenylmethane intermediates are too large for ZSM-5

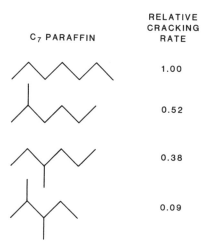

RELATIVE
CRACKING
C₇ PARAFFIN   RATE

1.00

0.52

0.38

0.09

Figure 4.26. Relative cracking rates of heptane isomers over ZSM-5.[140]

cavities, so disproportionation is spatially hindered and selectivity improved. Another example is aromatic formation from methanol. Although this reaction occurs readily over several acidic zeolites, only ZSM-5 limits product to high octane aromatics.[140]

When metal components are added through ion exchange, subsequent reduction gives new dimensions to these applications. Transition ions inside cavities are often difficult to reduce and are stabilized against agglomeration. This is an advantage, since extremely small, molecular-sized clusters of metal atoms may be produced, yielding activities and selectivities often found in organometallic homogeneous catalysis.[138] As more is known about the properties of these small metallic particles, many innovative uses will suggest themselves. Already, the principle of metal–zeolite combinations is commercially proven in hydrocracking, which combines both the features of hydrogenation and selective cracking to produce better gasoline components with less coking.[57]

Zeolites, with their vast possibilities of existing and potential structures, with shape and size selectivity, and with precise control of acid and other

TABLE 4.14. Relative Diffusivities of Xylenes in ZSM-5

Hydrocarbon	Relative diffusivity at 350°C
p-Xylene	1000
m-*o*,*m*-Xylene	1

TABLE 4.15. Major Industrial Zeolite Processes

Process	Zeolite	Products
Catalytic cracking	Faujasite	Gasoline, fuel oil
		Kerosene, jet fuel
Hydrocracking	Modified faujasite	Benzene, toluene,
		Xylene
Hydroisomerization	Mordenite	Gasoline and
		Distillates
		i-hexane, heptane
Dewaxing	ZSM-5 mordenite	Low power point lubes
Benzene alkylation	ZSM-5	Styrene, ethyl benzene
Xylene isomerization	ZSM-5	Paraxylene
Toluene		
disproportionation	ZSM-5	Benzene, toluene
Methanol-to-gasoline	ZSM-5	Gasoline
Methanol-to-olefins	Modified ZSM-5	Olefins
Fischer–Tropsch	Modified ZSM-5	Hycrocarbons

functions offer exciting materials for novel and innovative process applications.[142,135] This research includes not only the search for new processes but also investigation of other zeolite structures based on elements other than silicon, for example, phosphorus. Table 4.15 lists the major industrial processes using zeolites.

4.5. CONCLUDING REMARKS

This chapter has presented a summary of the important features of research on catalytic materials. It is a changing world and new data will be forthcoming. Nothing has been said about exciting new materials such as metal sulfates,[144] phosphates,[145] metal–graphite intercalation compounds,[146] superbasic materials,[147] and carbides or nitrides,[148] all of which await industrial exploitation.

DESIGN OF CATALYSTS
How to Invent Them

5.1. METHODOLOGY

Designing catalysts to satisfy process needs is like writing prescriptions to cure illnesses. We hope that the procedure is accurate and successful and try to base it on scientific method. Yet in practice, it is the skill of the practitioner, whose background, knowledge, experience, and awareness of "what has worked in the past" usually prevails. Nevertheless, in catalysis as well as medicine, we continually strive to improve the precision of the design process beyond mere copying of recipes.

All successful catalyst inventors have combined unique experience with existing knowledge, some very methodically, others with flashes of intuition and genius. It was only in the 1960s, however, that Dowden first attempted to systematize catalyst design so that scientifically trained yet inexperienced chemists and engineers could attempt this seemingly magical exercise.[149] Dowden's work has been extended by Trimm in a book devoted to the subject.[85] Trimm gives many examples, which are recommended as supplementary material to the discussion in this chapter.

The objective is to guide the knowledgeable novice through systematic procedures that lead to creative discoveries; a path seemingly easily traversed by the expert. Such an expert may appear intuitive, yet he himself passes through similar stages. With practice, the beginner will begin to inject his own style, to innovate, to optimize. Each case produces variations on the central theme, but yet the methodology is the same.

The procedure is attractive as a teaching vehicle. It has proven successful in many practical applications. As a training device, students progress faster into understanding the principles of catalyst design. If nothing more, the organization leads to a disciplined identification of critical features and structure of background material. Often, missing pieces in the puzzle of data suggest areas for further research.

Discussion on process development in Chapter 3 emphasized the importance of defining the process need and identifying the objectives to be achieved. If these call for diffusional or mechanical modifications, then the design will involve changes in preparation and formulation that optimize the particle size, pore structure, and strength. In cases where resistance to deactivation must be improved, the solution is found through incorporation of chemical or structural promoters, or perhaps by changing the shape of pores. Improvements in regeneration call for the incorporation of combustion additives.

In all these situations, catalyst design is essentially modification of existing catalyst composition and structure. The procedure may be difficult but is a fairly straightforward application of many principles discussed earlier. However, when an entirely new composition is required, the catalyst designer is faced with selecting suitable materials for further testing. The ease with which this is done depends on available research data. In the event that no proven process technology exists as a guideline, then the search must turn toward totally novel and untried materials. A suitable active component must be found. Other factors such as optimal activity, selectivity, lifetime, and formulation are subjects for future development.

A target reaction is essential. This is easiest with chemical processes since the critical reaction is easily identified. With petroleum and other fuel processes, so many reactions exist that some degree of lumping into reaction types, perhaps with model compounds, is necessary.

We now follow the original strategy of Dowden.[149] The distinct steps in the method are shown in Fig. 5.1. These are as follows:

1. Stoichiometric analysis.
2. Thermodynamic analysis.
3. Proposed molecular mechanism.
4. Proposed surface mechanism.
5. Reaction path identification.
6. Necessary catalyst properties.
7. Search for appropriate materials.
8. Proposed catalyst(s).

Each of these steps is demonstrated with an example, selected more for pedagogical purposes than practicality.

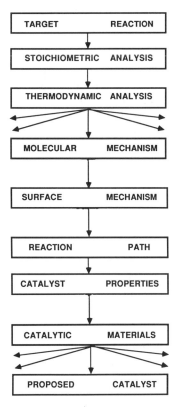

Figure 5.1. Steps in the catalyst design.

5.2. EXAMPLE—OXIDATION OF METHANE TO FORMALDEHYDE

The target reaction is the partial oxidation of methane to formaldehyde:

$$CH_4 + O_2 = CH_2O + H_2O$$

$$\Delta H^0_{r,25} = -76.8 \text{ kcal mole}^{-1} \qquad (5.1)$$

$$\Delta G^0_{r,25} = -70.9 \text{ kcal mole}^{-1}$$

This reaction is exothermic and thermodynamically feasible with large conversions possible even up to high temperatures.

Formaldehyde is now made from CH_4, but through a series of separate processes: steam reforming, methanol synthesis, and methanol oxidation. This complex process route leads to low efficiency and high cost. Direct partial oxidation to formaldehyde is a worthy process objective. We need a catalyst with high activity and selectivity for reaction (5.1).

5.2.1. Stoichiometric Analysis

Three tasks are involved in this step. First, all possible stoichiometric chemical equations are listed. These must be organized in a logical manner, for example, primary reactant, reactant self interactions, reactant cross interactions, reactant-products, and product-products. Only those involving chemically stable compounds are included.

Second, the thermodynamics for each reaction are calculated. The most useful property is $\Delta G_{r,T}^0$ at a specified temperature or range of temperatures. Computation of $\Delta H_{r,T}^0$ is also of value but not critical.

Third, types of chemical bond changes are identified for each reaction. These should be indexed according to some scheme, for example

Dehydrogenation	DH
Hydrogenation	H
Oxidation	O
Oxygen insertion	OI
Dehydration	DW
Group addition	A

Appendix 5 is an extensive compilation of reaction types and is useful in organizing a systematic nomenclature. For methane oxidation, we have, among others, the following reactions:

Primary Reactions:

$$CH_4 = none$$
$$O_2 = none$$

Self Interactions:

		$\Delta G_r^0(800\ K)$
$2CH_4 = C_2H_6 + H_2$	DH	8.5 kcal mole^{-1}
$\quad = C_2H_4 + 2H_2$	DH	12.8
$\quad = C_2H_2 + 3H_2$	DH	22.2
$2O_2 = none$		

Cross Interactions:

$CH_4 + \frac{1}{2}O_2 = CH_3OH$	OI	−20.6
$\quad = CH_2O + H_2$	OI, DH	−20.0
$\quad = CO + 2H_2$	OI, DH	−43.1

$$CH_4 + O_2 = CH_2O + H_2O \qquad \text{OI, DH, O} \qquad -70.9 \text{ kcal mole}^{-1}$$
$$= HCOOH + H_2 \qquad \text{OI, DH, O} \qquad -67.0$$
$$= CO + H_2 + H_2O \qquad \text{OI, DH, O} \qquad -87.3$$
$$= CO_2 + 2H_2 \qquad \text{OI, DH, O} \qquad -90.5$$
$$CH_4 + \tfrac{3}{2}O_2 = CH_2O + H_2O_2 \qquad \text{OI, DH, O} \qquad -31.0$$
$$= HCOOH + H_2O \qquad \text{OI, DH, O} \qquad -119.8$$
$$= CO + 2H_2O \qquad \text{OI, DH, O} \qquad -136.5$$
$$= CO_2 + H_2 + H_2O \qquad \text{OI, CH, O} \qquad -139.8$$
$$CH_4 + 2O_2 = HCOOH + H_2O_2 \qquad \text{OI, DH, O} \qquad -98.6$$
$$= CO + H_2O_2 + H_2O \qquad \text{OI, DH, O} \qquad -118.7$$
$$= CO_2 + 2H_2O \qquad \text{OI, DH, O} \qquad -189.5$$

Reactant–Product Reactions:

$$CH_4 + C_2H_6 = C_3H_8 + H_2 \qquad \text{DH, A} \qquad 16.6$$
$$CH_4 + C_2H_4 = C_3H_8 \qquad \text{A} \qquad 4.5$$
$$CH_4 + CH_3OH = C_2H_5OH + H_2 \qquad \text{DH, A} \qquad 10.5$$

$$\tfrac{1}{2}O_2 + H_2 = H_2O \qquad \text{OI} \qquad -50.3$$
$$\tfrac{1}{2}O_2 + C_2H_6 = C_2H_5OH \qquad \text{OI} \qquad -30.7$$
$$\tfrac{1}{2}O_2 + CH_2O = HCOOH \qquad \text{OI} \qquad -45.3$$

$$O_2 + H_2 = H_2O_2 \qquad \text{O} \qquad -9.2$$

5.2.2. Thermodynamic Analysis

The objectives in this step are to assess thermodynamic feasibility for each of the stoichiometric equations and to list them in groups with the same chemical functions. For clarity, many intermediate reactions have been omitted from the rearrangements below. Reactions with ΔG_r^0 values greater than 10 kcal mole^{-1} are not included at this point.

$$CH_4 + 2O_2 = CO_2 + 2H_2O \qquad \text{OI, DH, O} \qquad -189$$
$$CH_4 + O_2 = CH_2O + H_2O \qquad \text{OI, DH, O} \qquad -70$$
$$CH_4 + O_2 = HCOOH + H_2 \qquad \text{OI, DH} \qquad -67$$
$$CH_4 + \tfrac{1}{2}O_2 = CH_2O + H_2 \qquad \text{OI, DH} \qquad -20$$
$$CH_4 + \tfrac{1}{2}O_2 = CH_3OH \qquad \text{OI} \qquad -22$$
$$CH_2O = CO + H_2 \qquad \text{DH} \qquad -17$$
$$CH_3OH = CH_2O + H_2 \qquad \text{DH} \qquad 2$$

At this stage, the reactions are examined for common trends. For example, the target reaction is in the same grouping as combustion or oxidation. Attempts to build these functions into the catalyst result in poor selectivity. Searching further, the analysis shows a possible alternative reaction

$$CH_4 + \tfrac{1}{2}O_2 = CH_2O + H_2 \qquad (5.2)$$

involving only oxygen insertion and dehydrogenation. This may proceed directly or involve a combination of the next reaction, with CH_3OH as an intermediate that decomposes via the last reaction. At this point, one possibility is to focus upon equation (5.2) as the target.

5.2.3. Proposed Molecular Mechanism

The purpose of this step is to visualize the molecular events. In this case, the path is fairly simple, as shown in Fig. 5.2. However, other examples may have any number of proposed molecular mechanisms, with each one a starting point for an analysis to follow. Some discretion is needed to differentiate among competing possibilities.

5.2.4. Proposed Surface Mechanism

This is the point where prior experience of the designer and some knowledge of existing surface chemistry is invaluable. Obviously the surface mechanism is just a guess, but it should be as informed as possible. There will be many possible surface schemes, each one leading to possible catalysts. If time permits, a number of these could be investigated. Otherwise, some priorities must be assigned, based on as much information as possible. Exercises of this type, although speculative, often have side benefits since

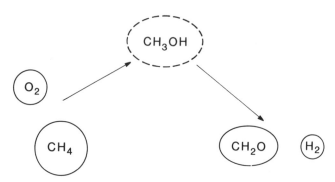

Figure 5.2. Proposed molecular mechanism.

gaps in existing data become apparent and lead to ideas for future research. For this example, the mechanism shown in Fig. 5.3 was adopted as a "best guess." The reader should appreciate that others are possible.

Both CH_4 and O_2 are chemisorbed and dissociated. It is known that CH_4 may adsorb as CH_x with x from 0 to 3.[150] Excessive dehydrogenation is undesirable since CH_2 is much more reactive than CH_3 and is likely to lead to oxidative products. Also, strong dehydrogenation could also lead to decomposition of the product CH_2O, as it forms.

Oxygen adsorbs as molecular ionic and atomic species.[151] The latter are preferred since the product contains only one oxygen. However, adsorbed oxygen atoms are reactive and lead to oxidation. There are correlations between combustion activity and mobility of these surface species.[152] To avoid oxidation, we require that oxygen dissociation leads to chemisorbed bonds that are too strong to allow migration. This means that the adsorbed CH_3 must be sufficiently mobile to diffuse to oxygen sites. Reaction between O and CH_3 follows, and the intermediate dehydrogenated to give formaldehyde.

5.2.5. Reaction Path Identification

The necessary reaction paths have now been established. The catalyst must promote oxygen insertion and mild dehydrogenation but inhibit strong oxidation and dehydrogenation.

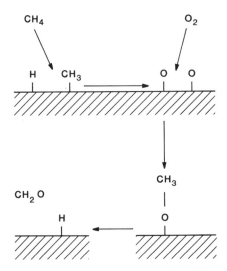

Figure 5.3. Surface mechanism for formaldehyde formation.

5.2.6. Necessary Catalyst Properties

This step is merely a redefinition of the surface mechanism in terms of properties helpful in the search for appropriate materials. The catalyst must have

1. Oxygen adsorption sites, leading to dissociation and immobile oxygen species.
2. Mild dehydrogenation sites that give CH_3 from CH_4 and remove H from the intermediate.
3. Adjacent sites in order to facilitate the final dehydrogenation step.

5.2.7. Search for Appropriate Materials

The requirement for mild dehydrogenation and low oxidation eliminates most metals, focusing the search upon oxides. Figure 5.4 shows a typical pattern of activity for dehydrogenation.[45] High activity is centered around ions with d^4 and d^6 configurations. Mild dehydrogenation is found in oxides containing ions such as Cu^{2+}, Ni^{2+}, Fe^{3+}, Mn^{2+}, V^{3+}, V^{5+}, and Ti^{4+}.

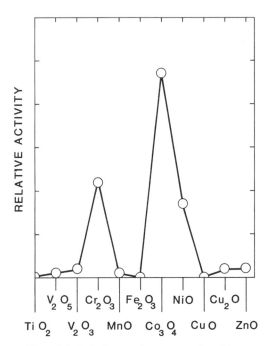

Figure 5.4. Dehydrogenation patterns in oxides.

Oxygen insertion activity in oxides is essentially mild oxidation. Clues for potential materials are found from three sources. The first is a pattern of behavior for CO oxidation, giving[153]

Strong oxidation: Co^{2+}, Ni^{2+}, Cu^{2+}, Pd^{4+}, Ag^{1+}, Cd^{1+}, Mn^{2+}
Moderate oxidation: Rh^{4+}, In^{3+}, Sn^{3+}, Pb^{2+}
Mild oxidation: Sc^{3+}, Ti^{4+}, V^{3+}, Cr^{3+}, Fe^{2+}, Zn^{2+}, Zr^{3+}, Nb^{3+}, Mo^{6+}

Second, oxygen mobility correlates well with oxidation.[152] Available measurements give the sequence

$$Co_3O_4 > MnO_2 > NiO > CuO > Fe_2O_3 > Cr_2O_3 > V_2O_5 > MoO_3$$

Third, activity for strong oxidation is found in oxides that reduce easily, suggesting that reduction followed by oxidation may be a mechanism.[45] Oxides that are hard to reduce include

Aluminates: $CoAl_2O_4$, $NiAl_2O_4$
Titanates: $ZnTiO_4$

5.2.8. Proposed Catalyst

By excluding strong oxidation ions from the mild dehydrogenation list, we arrive at the following listing in order of decreasing activity:

Mild dehydrogenation	Oxygen insertion
Fe^{3+}	$Sc^{3+}V^{3+}$
V^{3+}	Ti^{4+}
V^{5+}	V^{3+}
Ti^{4+}	Fe^{2+}
	Zn^{2+}
	Zr^{3+}
	Nb^{3+}
	Mo^{6+}

Candidates that are consistent with all these properties are

Single oxides: TiO_2, V_2O_3
Mixed oxides: $TiO_2 + MoO_3$, $ZnO + V_2O_3$
Complex oxides: Fe_3O_4, Fe molybdate, Zn TiO_3

Dispersion will not be necessary, since high areas only facilitate complete oxidation. Therefore, simple methods of preparation will be sufficient. These materials are suggestions only. No priorities can be assigned and each must be tested. Only then may the next task of optimization and modification proceed.

5.3. COMPUTER AIDS

Knowledgeable readers will have no doubt perceived that this procedure has two obvious faults. A critical point in the analysis is selection of the surface mechanism, where the specialized knowledge of the designer is important. Thus expertise has not yet been completely eliminated. One individual may have better success than another. Second, the process of writing chemical reactions, calculating thermodynamics, and so on is tedious work. Even in the example, we restricted calculations to one temperature for convenience. Third, both the number of reactions and possible mechanistic paths appear endless. Listing of stoichiometric equations, for example, is limited by the patience of the designer. Many molecular mechanisms are feasible, each leading to multiple surface mechanisms, which in turn generate a number of candidate catalysts. Although the expertise of the designer may help eliminate unlikely possibilities and even assign priorities to others, there is a danger that impatience, preconceptions, and prejudice may lead to interesting avenues being overlooked. If the task is performed properly, it is too time consuming; if not, we risk loss of possibly successful reaction paths through omission. A solution is to use computer-aided catalyst design.

Modern computers have enormous capacity to store data and perform analysis. Stoichiometric analysis and thermodynamic interpretation are easily programmed. The range of reactions considered is thus expanded. Also, correlations and patterns of behavior lend themselves to organization. In this way, the accumulated background of many experts is at the disposal of the designer. Even the process of mechanism generation is not beyond computer capabilities. As with computer-aided process design, many possibilities could be developed, priorities assigned, and discrimination left to the human designer, who, free from manual tedium, is now able to let his imagination and intuition roam. Figure 5.5 shows the structure of the computer-aided design. The most effort is needed in software development and data base organization.

5.4. SOME FINAL THOUGHTS

Before the reader concludes that the catalyst expert is vanishing and that future inventions will come from the uninitiated technologist or even computers, it should be pointed out that much remains to be done before catalysis is truly a science. There is still a need for the alchemist in us all. But the shrouds of mystery and magic are falling. The beginner should no longer fear to travel with the mighty.

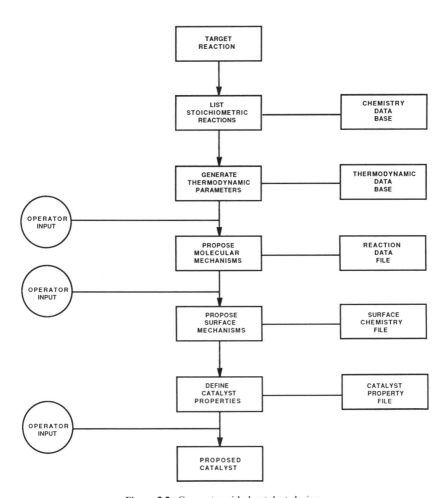

Figure 5.5. Computer-aided catalyst design.

CATALYST PREPARATION
How They Are Made

6.1. INTRODUCTION

Although some degree of catalytic activity is found in any given material, successful preparations proceed through a series of mysterious steps justifying the judgment "black art." Indeed, it was this very fact that prompted the following quotation from Raney[154]:

> It is in the preparation of catalysts that the chemist is most likely to revert to type and to employ alchemical methods. The work should be approached with humility and supplication and the production of a good catalyst received with rejoicing and thanksgiving.

In this chapter, we find that catalyst preparational methods are not always magical but that some of the elements mentioned by Raney are indeed necessary for success. Catalysts are considered in the order listed in Table 6.1.

6.2. SINGLE ACTIVE COMPONENTS AND SUPPORTS

Single oxide active components and supports are usually prepared with high surface area and porosity. Similar techniques apply in the production of both, but the major use of these materials is for supports.[53] Procedures used are those found in the preparation of colloidal hydrous oxides, with variations inherent to each type.[155] The preparational steps are given in Fig. 6.1.

95

TABLE 6.1. Types of Catalysts

Category	Class	Examples
A. Single active component and supports	1. Single oxides	Al_2O_3, SiO_2, Cr_2O_3
	2. Dual oxides	SiO_2-Al_2O_3, NiO-Al_2O_3, CuO-Al_2O_3 zeolites
B. Deposition-produced activity component	3. Dispersed oxides	MoO_3/Al_2O_3
	4. Dispersed metals low loading	0.3% Pt/Al_2O_3
	5. Dispersed metals moderate loading	40% Ni/Al_2O_3
C. Extraction-produced active component	6. Dispersed metals high loading	70% Ni/Al_2O_3
	7. Porous metals	Raney Ni
	8. Fused oxides	Fe_3O_4-Al_2O_3-K_2O
D. Special types	9. Mixed oxides	ZnO, $ZnCr_2O_4$
	10. Cemented oxides	NiO-$CaAl_2O_4$
	11. Metal gauze	Pt, Ag

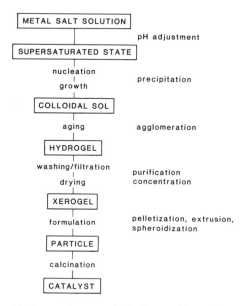

Figure 6.1. Preparation steps for hydrous oxides and supports.

6.2.1. Metal Salt Solution

The first step is to prepare a solution (water is the preferred solvent) of a metal salt, M_nX_m, destined to become the oxide M_xO_y. The solubility of the salt must be sufficient to give convenient volumes at prescribed temperatures. If necessary, organic solvents are used. The amount of solvent is determined by the quantity of oxide desired, size of laboratory vessels, and requirements of other steps in the preparation. Choice of the anion is based on many factors, such as solubility, impurities, availability, cost, and potential problems.

The last item refers to the difficulty often encountered in removing adsorbed anions from the precipitated oxide. As we shall see, the nature of X (e.g., Cl^-, NO_3^-, or SO_4^{2-}) influences the stability of the precipitate. A certain amount adsorbs on the particles. These must be removed, either by washing or volatilization during drying and calcination. Chlorides left on the catalyst increase acidity, sulfates form either SO_2 or H_2S, depending on conditions, and deactivate other components. Nitrates produce obnoxious fumes during calcination. Certain compromises are necessary, for example, oxalates are the best but are not always readily available and sometimes evolve toxic compounds upon calcination. Sulfates are the least expensive but are difficult to remove.

6.2.2. Controlled Precipitation

The objective of this step is to precipitate a sol, a colloidal particle $10–10^3$ nm in diameter. Sol particles do not settle, are difficult to filter, and are not visible except with an ultra microscope. They are the beginning of the process leading to the formation of porous structure in the catalyst. If precipitation is too vigorous, then massive particles are formed and lack the necessary properties for high surface catalysts.

Precipitation occurs in three phases: supersaturation, nucleation, and growth. Pertinent parameters producing supersaturation are shown in Fig. 6.2. Solubility curves are a function of temperature and pH. In the supersaturated region the system is unstable and precipitation occurs with any small disturbance. Precipitation can be rapid and agglomeration severe. Slow growth is possible in the metastable solutions, but only if unfavorable conditions are avoided. This metastable supersaturation region is approached either by increasing the concentration through evaporation (A to C), lowering the temperature (A to B), or increasing the pH (which effectively moves the solubility curve to D and A into the supersaturation region). This last approach is the most convenient method. The reaction

$$M^{n+} + nOH^- \leftrightarrows M(OH)_n \tag{6.1}$$

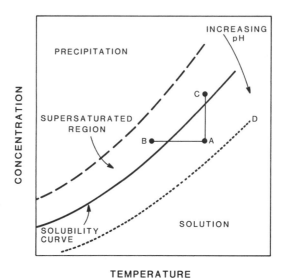

Figure 6.2. Metastable, supersaturated states.

is controlled by increasing the pH through addition of a basic solution. Commonly used reagents are sodium, potassium, and ammonium hydroxides, carbonates, and bicarbonates. The choice of which one to use is determined by subsequent action of the associated cations. Table 6.2 gives the precipitation conditions for common transition hydroxides.

Particles within the supersaturated region develop in a two-step process; nucleation and growth. Nucleation may proceed spontaneously through the formation of $M(OH)_n$ entities or be initiated with "seed" materials. These are solid impurities such as dust, particle fragments, or rough edges on

Table 6.2. Precipitation pH for Hydrous Oxides $(25°C)^a$

Hydroxide	pH
$Mg(OH)_2$	10.5
$Mn(OH)_2$	8.6
$Co(OH)_2$	6.8
$Ni(OH)_2$	6.7
$Fe(OH)_2$	5.5
$Cu(OH)_2$	5.3
$Cr(OH)_3$	5.3
$Zn(OH)_2$	5.2
$Al(OH)_3$	4.1
$Fe(OH)_3$	2.0

^{a} Reference 157.

vessel surfaces. The rate of nucleation may be accelerated by deliberate addition of seed nuclei. Growth then proceeds through adsorption of ions on the surface of the seeded particle. This growth process proceeds at a rate that is a function of concentration, temperature, and pH.

Thus the relative rates of nucleation and growth may be independently controlled to some extent. If nucleation is faster than growth, the system produces a narrow distribution of small particles. Fast growth results in narrow distributions of larger particles. When the rates are similar, wide size distributions result.

Precipitation of sols may also be initiated by condensation, by hydrolysis of organic soluble complexes, and by oxidation or reduction.[156]

6.2.3. Agglomeration and Gelation

Sols have large surface energies which the system strives to minimize through particle growth or agglomeration.[156] There are three types of agglomerates: hydrogels, flocks, and massive precipitates. Hydrogels are the most important in catalysis.

Hydrogels have a three-dimensional, loosely bound structure, as shown in Fig. 6.3. Small particles set up long-range order through hydrogen bonding via the interstitial water molecules. Better long-range homogeneity is achieved with gels derived from sols with a narrow size distribution. However, these particles are mobile and grow through a process of collision and coalescence. In order to appreciate factors controlling this growth, it is necessary to examine the nature of interaction between these hydrous oxides and water molecules,

$$
\begin{array}{c}
\quad\ \ \text{OH}\quad\ \ \text{OH}\qquad\ \ \text{O}^- \\
\quad\ \ | \qquad\ \ | \qquad\quad | \\
\text{H}_2\text{O} + -\text{M}-\text{O}-\text{M}- \ \rightleftarrows\ -\text{M}- + \text{H}_3^+\text{O}
\end{array}
\qquad (6.2)
$$

$$
\begin{array}{c}
\quad\ \ \text{OH}\quad\ \ \text{OH}\qquad\ \text{OH}_2^+ \\
\quad\ \ | \qquad\ \ | \qquad\quad | \\
\text{H}_2\text{O} + -\text{M}-\text{O}-\text{M}- \ \rightleftarrows\ -\text{M}- + \text{OH}^-
\end{array}
$$

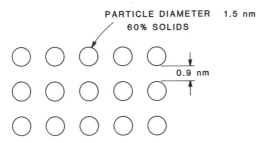

Figure 6.3. Structure of a hydrogel.

The surface may be positively or negatively charged depending on the pH. With low pH acid solutions, the equilibrium is driven toward positive surfaces. The surface becomes less positive and finally negative as the pH increases. However, the effective charge on the surface is partially neutralized by counterions in the solution. In Fig. 6.4, a positively charged particle attracts anions in the solution. These originate from the bases used during precipitation or may be electrolytes added during aging. Counterions form a space charge, part of which is held sufficiently strongly to be carried along as the particle moves with Brownian motion. The result is an effective charge, called the zeta potential. Both the original charge and the neutralizing counterions respond to pH changes, resulting in the zeta potential curves shown in Fig. 6.5.

The zeta potential determines the rate of gelation. If the charge is high, particles effectively repel one another and avoid contact. If it is low, then thermal motion leads to collision and coalescence. These rates are highest at the isoelectronic point, where the zeta potential is zero.

These effects are demonstrated in Fig. 6.6 showing surface areas and gelation times for silica sols.[158] From a practical point of view, it is necessary to compromise between large surface areas and reasonable gelation times.

Silica gels, such as those in Fig. 6.6, are prepared by mixing solutions of water glass (sodium silicate with $SiO_2/Na_2O = 3.22$) and HCl. This is an example of condensation and produces $(HO)_3Si-O-Si(OH)_3$ sols of about 1.5 nm at a pH of 6. Gelation times of 10 min result in gels so stiff they may be cut into cubes. The hydrogel has a pore volume of 2.0 cm^3 g^{-1} and contains about 60%–70% H$_2$O.

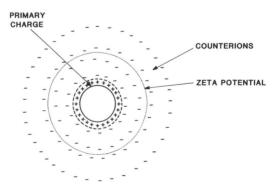

Figure 6.4. Double layer structure of a charged particle.

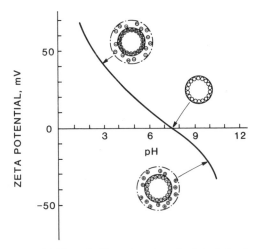

Figure 6.5. Effect of pH on the zeta potential of alumina.

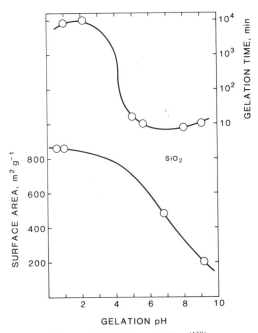

Figure 6.6. Gelation of silica.[158]

Promoters, if needed, may be added during the gelation period. Alternatively, they may be deposited, using the same methods described later for active components.

6.2.4. Washing and Filtering

The next two steps, washing and filtering, are sometimes reversed. A satisfactory, but time-consuming, method is to wash by decantation. The hydrogel is added to a large volume of distilled water in a conveniently sized container and the suspension thoroughly mixed. Upon standing, particles settle slowly, desorbing foreign ions as they fall. When a definite interface is visible, water is removed by decantation and the process repeated. The purified gel takes longer to settle with each washing, since removal of counterions increases the zeta potential. The gel partly reverts back to a sol, a process known as peptization. Care must be taken not to wash too much or settling times become too great. A good method is to check impurity levels in the decantation water during the operation. When washing is complete, the hydrogel is filtered.

If filtering is done first, the filtrate is washed with successive amounts of water until a satisfactory level of impurity in the wash water is found. Again, care must be taken not to wash too much or the gel peptizes and passes through the filter paper.

6.2.5. Drying

Drying is necessary in order to remove the large volume of water in the hydrogel. Some collapse of the structure is to be expected, as shown in Fig. 6.7, but care must be taken to properly control drying operations if high porosity is desired.

Initially, drying occurs through evaporation of moisture from the outside surface of the hydrogel. The rate of water loss is constant and the mass transfer controlled by temperature, relative humidity, flow rate of air over the surface, and size of the filtrate. This process continues until moisture content drops to about 50%. The filtrate mass now begins to shrink as most of the external water disappears. The material is now called a xerogel.

Continued moisture loss occurs with a declining rate, in which evaporation is controlled by capillary forces. The saturation point decreases as pores become smaller and evaporation slows until water is forced into larger pores by concentration gradients. If evaporation occurs but removal of moisture is blocked by smaller pores, then large internal pressures of steam

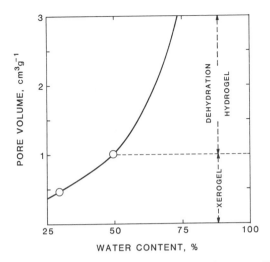

Figure 6.7. Pore volume loss on drying of a silica hydrogel.[158]

develop and the structure collapses, with loss of pore volume and surface area. High temperature gradients in the sample must be avoided. The effect of temperature is shown in Fig. 6.8. Lower temperatures give less surface loss, since evaporation rates are lower.[159] Vacuum drying at lower temperatures is a satisfactory laboratory procedure. In fact, one of the best devices is a rotary, lamp-heated evaporator.

Control of evaporation rates is much easier with large-scale equipment. Continuous operations such as belts, rotary kilns, fluidized, and spray driers

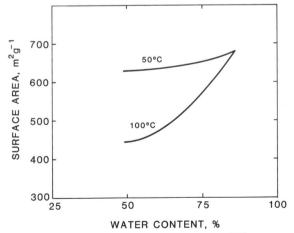

Figure 6.8. Temperature effects in drying.[158]

give more consistent and uniform results than laboratory ovens, hot plates, and furnaces.[35,160] Often these units are staged to achieve optimum results, for example, rapid, humid drying during the constant rate period, followed by slow control during the declining rate.[161]

Dried xerogels contain 25%–30% water, encapsulated in fine pores or chemically bound to the oxide. In this moist state the material is sometimes easier to form into pellets and extrudates. This is sometimes done for convenience, provided subsequent calcination does not weaken or otherwise harm the particles.

6.2.6. Calcination

Calcination is further heat-treatment beyond drying. Several processes occur: (1) loss of chemically bound water or CO_2, (2) changes in pore size distribution, (3) active phase generation, (4) surface conditioning, and (5) stabilization of mechanical properties.

Alumina is a good example with which to demonstrate all of these features.[64] Figure 6.9 shows the progression through all steps from precipitation to calcining. After drying, a hydrous oxide known as "boehmite" is produced with a structure $Al_2O_3 n H_2O$. With pure boehmite $n = 1$, but values up to 1.8 are found. The structure of boehmite is distinctive and the surface hydrated as shown in Fig. 4.17. Upon calcination above 300°C, a series of

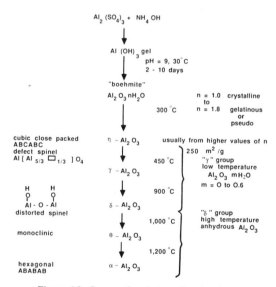

Figure 6.9. Preparational steps for aluminas.

phase changes occur, simultaneously with loss of hydroxyl groups. This results in η-, γ-, and δ-phases approximately at the temperatures shown. Exact transition points depend on many preparational factors and are difficult to define. These oxides are known as the "γ-group" or "low-temperature" aluminas. Structures are very similar, with a formula $Al_2O_3 n H_2O$, in which $n = 0.6$-0. All are based on cubic close packing of oxygen ions (ABCABC) but are best described as a defect spinel, $Al[Al_{5/3}[\quad]_{1/3}]O_4$, where Al occupies tetrahedral or octahedral interstices and [ ] is a cation vacancy. Different phases are distinguished by variations in relative intensities of key diffraction lines.

Above 1000°C, monoclinic θ-Al_2O_3 forms, transforming to hexagonal (ABABAB) α-Al_2O_3 at 1200°C. These are anhydrous, low surface area oxides and are not suitable for porous supports. They are used in applications where mechanical strength, and not high surface area, is required.

The acidity of these aluminas was discussed in Chapter 4. In particular, the effect of calcination temperature on acid strength distribution is shown in Fig. 4.18. Collapse of smaller pores results in increasing pore size, as shown in Fig. 6.10. Changes in mechanical properties come from subsequent differences in, for example, plasticity or grain boundaries. These factors are important during pelleting and extrusion.

Common supports such as Al_2O_3 and SiO_2 are made by catalyst manufacturers using modifications of these procedures. Excellent product quality control is achieved with a wide range of properties. In preparing catalysts, the laboratory chemist is advised to select suitable supports from among those available, unless changes in procedure are essential. Much labor, time, and frustration is saved. Suppliers of supports are listed in

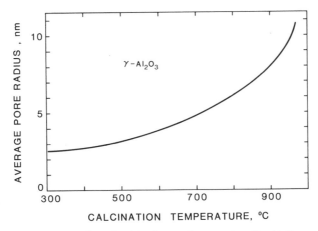

Figure 6.10. Effect of calcination on the pore size of γ-Al_2O_3.

Appendix 7. For do-it-yourself enthusiasts, detailed recipes are given in Appendix 6.

6.3. DUAL OXIDES

Dual oxides are intimate combinations of two oxides, resulting in synergistic catalytic properties. For high dispersion and interaction, best results are obtained through coprecipitation and subsequent treatment of the two gels. Common examples are $SiO_2-Al_2O_3$, $NiO-Al_2O_3$, and zeolites. Dual oxides are not separate compounds, since they exist over a wide range of compositions, but are more like solid solutions. This is especially true for low calcination temperatures. With high heat treatments, solid state reactions occur and identifiable compounds "precipitate" out.

6.3.1. $SiO-Al_2O_3$

The best known dual oxide is $SiO_2-Al_2O_3$, used as a cracking catalyst or acidic support. The preparation is very similar to that described for SiO_2. Two solutions are prepared, the first a 30% mixture of water glass and water, the second a mixture of $4N$ HCl and $0.5N$ $Al_2(SO_4)_3$. These are cooled to 5°C and mixed rapidly in a 2:1 ratio. As the pH of the acidified water glass reaches 6, precipitation of both silica and alumina sols begin and gelation takes about 45 sec. The combined gel is aged for two days, treated with 2% $Al_2(SO_4)_3$ solution to ion-exchange Na, and washed free of sulfate ions. Preparation is completed with drying and calcination at 550°C.[163]

The resulting catalyst contains about 25% Al_2O_3 and is amorphous, i.e., has no distinctive x-ray diffraction pattern indicating long-range crystallinity. Acidic properties are described in Chapter 4.

6.3.2. $NiO-Al_2O_3$

Coprecipitated $NiO-Al_2O_3$ samples are typical of a wide variety of dual hydrous oxide gels. They are precursors to reduced Ni/Al_2O_3 catalysts used in hydrogenation and methanation. Although these catalysts are produced by other routes, the intimate interactions fostered by coprecipitation are believed to impart improvements in activity and stability to the nickel crystallites,[164] although mechanical strength is lower. It may also be possible that irreducible compounds of NiO and Al_2O_3, facilitated by coprecipitation, play some role in forming stable promoters.

Both nickel and aluminum precipitate hydrous oxide sols. For best combinations, care must be taken to ensure that precipitation occurs simultaneously. Variations of pH within the vessel or during addition of the alkali result in preferential precipitation of one component, yielding product nonuniformity. One laboratory approach is to use the apparatus shown in Fig. 6.11.[165]

The inner flask contains a solution of $6N$ NaOH, the outer appropriate mixtures of nickel and aluminum nitrates. Tests are made to find the correct ratio between the two solutions such that, after mixing and precipitation, the final pH is 7. Inverting and shaking the jar gives instantaneous coprecipitation.

When sodium bicarbonate is used as the precipitating agent, a distinct nickel, aluminum hydroxy carbonate compound is formed.[163] This helps ensure the homogeneity desired in this type of material. Washing is best achieved with decantation. Usually 5–6 washes are sufficient. After filtering, the light green hydrous oxide is dried, crushed, and calcined. Calcination of coprecipitated oxides initiates solid state reactions in addition to those given for single oxides. In this case

$$NiO + Al_2O_3 = NiAl_2O_4 \qquad (6.3)$$

Nickel aluminate is a well-defined spinel that is difficult to reduce. Formation of pseudospinels are not limited, however, to stoichiometric

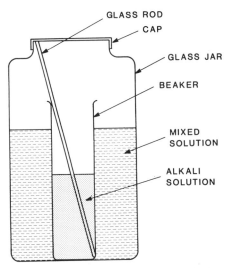

Figure 6.11. Apparatus for rapid coprecipitation.

compounds. This is because the defect spinel structure of γ-Al_2O_3

$$Al[Al_{5/3}[\ ]_{1/3}]O_4 \qquad (6.4)$$

favors solid solution with transition oxides such as FeO, CoO, NiO, and CuO. Cations, M^{2+}, do not necessarily occupy the octahedral vacancies but migrate to positions with higher stabilization energy. Thus we find $Al[AlFe]O_4$, $Co[Al_2]O_4$, $Al[AlNi]O_4$, and $Cu[Al_2]O_4$. The cations occupy stable positions and are difficult to reduce. These same procedures are used to prepare catalysts for low temperature shift and alcohol synthesis reactions, $CuO-ZnO-Al_2O_3$. In this case, mixed gels of $CuO-Al_2O_3$ are easily precipitated, with ZnO added at this or some later point.[166]

Coprecipitated oxides are the precursors to metal catalysts in Table 6.1 that are produced by extraction.

6.3.3. Zeolites

Zeolite synthesis was first attempted by duplicating the hydrothermal genesis of naturally occurring minerals. Mostly these were unsuccessful. Breakthroughs came with the realization that cogels of aluminosilicates, like those in $SiO-Al_2O_3$ preparations, are depolymerized by OH^- ions to yield crystalline nucleii. Depending upon composition, temperature, pressure, and pH, these nucleii develop into a vast number of zeolitic types.[133,136] With correct conditions for supersaturation, crystalline seeds grow into microcrystalline powders of appropriate material.[35] For example, Faujasite Y is made by a procedure similar to that for silica-alumina. Sodium silicate and sodium aluminate solutions in the correct proportions ($SiO_2/Al_2O_3 = 3-30$) are mixed, not with an acid, but with sodium hydroxide at a pH of 12. Finely divided silica is added as a seeding agent, and the temperature raised from 100 to 400°C, with pressure necessary at higher temperatures. Nucleation and crystallization commence at rates determined by the temperature and pressure. This digestion continues until the reaction is complete, taking a period of time that may vary from hours to days.[167] Other zeolites are made in similar ways, with essential differences in starting conditions and compositions.[138]

6.4. DEPOSITION OF THE ACTIVE COMPONENTS

Dispersion of oxides on high-area supports is carried out by one of four methods: (1) precipitation, (2) adsorption, (3) ion exchange, and (4) impregnation. Each technique has advantages and disadvantages. Often preference for one method over another is a matter of compromise. The supports are either in powder or particle form. Depositing active components on the internal surface of porous particles requires special attention to avoid pore diffusional limitations that give uneven distributions. This must be weighed against the advantages of better control of pore size distribution and easier handling operations. Active components deposited in large concentrations onto powders change mechanical and surface properties, so that subsequent pelleting or extrusion may be more difficult. In laboratory preparations, deposition onto powders is usually practiced. Commercial manufacturers find preformulation more convenient, efficient, and economical.

6.4.1. Precipitation

In precipitation, the objective is to achieve a reaction of the type

$$
\begin{array}{c}
\begin{array}{ccc}
\text{Metal salt solution} & + & \text{Support} \\
\text{oxalate} & & \text{powder} \\
\text{nitrate} & & \text{particle} \\
\text{sulfate} & & \\
\text{chloride} & &
\end{array}
\xrightarrow[\substack{\text{NaOh}\\ \text{KOH} \\ \text{NH}_4\text{OH} \\ \text{Na}_2\text{CO}_3 \\ \text{NaHCO}_3 \\ \vdots}]{\text{Base}}
\begin{array}{c}
\text{Metal hydroxide} \\
\text{or carbonate} \\
\text{on support}
\end{array}
\end{array}
\tag{6.5}
$$

The choice of salt or alkali depends on factors similar to those considered for single oxide precipitations, the most important being the possible harmful effects on the final catalyst. Powders or particles are slurred with an amount of salt solution sufficient to give the required loading. It is helpful to carry out preliminary heating or evacuation to ensure that the pores are properly filled with solution. The sequence of events is shown in Fig. 6.12.

Enough alkali solution is added to cause precipitation, the powder is then filtered or otherwise separated and washed to remove alkali ions, reagent anions, and excess deposit on the outside of particles. Two processes are involved in the deposition: (1) precipitation of sols in bulk and pore fluid, and (2) interaction with the support surface. Best results occur when the OH groups of the support surface enter into reaction (6.5) so that the pH of the surface region is higher than in the bulk solution. Precipitation then occurs preferentially and uniformly on the surface. For example, silanol groups of silica result in surface deposits of hydrosilicates rather than hydroxides.[168] The effect is not as pronounced with alumina, where the surface acts more like nucleating centers for sol formation. Often the process is merely one of sol adsorption or impregnation.

Rapid nucleation and growth in the bulk solution must be avoided, since sols are then too large to enter the pores easily and associate only with the outside of the particle. This is most likely to occur in the vicinity of the alkali droplets entering the solution. Rapid mixing is essential.

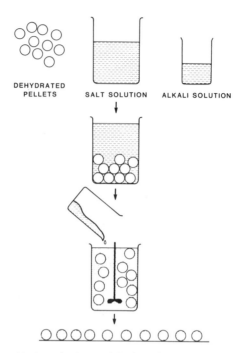

Figure 6.12. Steps in the precipitation of the active component.

Sometimes cooling, which lowers nucleation and growth rates, is effective. At other times, boiling the solution to introduce turbulence seems to work. Usually, mixing is no problem with laboratory devices. Dilute alkali may be added, drop by drop, with rapid agitation to disperse the droplets before local concentrations become excessive. Alternatively, both the salt and alkali solution may be added simultaneously in a controlled manner to a well-mixed container of water. Mixing is more difficult in commercial operations. Vessels are larger and mixing of reagents takes longer. This is one of the many perplexing problems faced during scale-up.[35]

One effective solution to control uniform precipitation is to use urea rather than conventional akalis. Urea dissolves in water but decomposes very slowly above 90°C. Appropriate amounts of urea are added to the metal salt-support slurry and the mixture heated with stirring. At 90°C, urea hydrolyzes and OH groups are formed uniformly throughout the vessel and in the pores. Precipitation takes place homogeneously over the surface. Since hydrolysis is slow and precipitation rapid, OH groups are consumed as fast as they are formed and the pH of the solution remains unchanged. Although requiring longer times, this technique yields very uniform products. Loading is controlled by the time of reaction. Scale-up also poses no

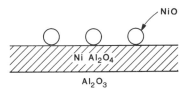

Figure 6.13. Phases of nickel compounds present after calcination of nickel alumina catalysts.

difficulties.[168] Variations of this procedure are practiced in which alkali solution is slowly injected through a hyperdermic syringe into the turbulence of the mixer.

After washing, the treated support is dried to remove excess moisture from the pores. This operation is not as critical as in support preparation, since the active component is firmly anchored to the surface. However, precautions should be taken to avoid rapid heating that generates large internal steam pressures. Calcination decomposes the deposited hydroxides and carbonates into stable oxides or metals, depending on the atmosphere. Temperatures are determined by the conditions necessary for decomposition. For example, nickel hydroxide deposited on alumina decomposes to the oxide at 300°C, whereas nickel hydrosilicate or silicate is stable up to 600°C.[168] Side reactions also occur, either parallel or series, between the original deposit, the oxide, and the support. Unlike coprecipitated systems, interaction is restricted to surface layers and may be only a few atomic dimensions thick. For example, the surface of NiO/Al_2O_3, prepared through precipitation and calculation, is shown in Fig. 6.13.

Whether $NiAl_2O_4$-type substrates are undesirable because they rob the system of NiO or whether they contribute some benefit, for example by stabilization during reduction, is a debatable point. The relative amounts of NiO and $Ni[Al_2]O_4$ depend on the calcination temperature, which should exceed any anticipated process temperatures.

Precipitation is the preferred deposition route for loadings higher than 10%–20%. Below this value, other techniques are usually practiced.

6.4.2. Adsorption

Support materials exposed to metal salt solutions adsorb equilibrium quantities of salt ions and obey adsorption isotherms as shown in Fig. 6.14. Adsorption is an excellent method for depositing small amounts. Powders or particles are dehydrated and "soaked" in the appropriate solution for suitable periods. Deposition is uniform, providing all pores are penetrated during the soaking time.

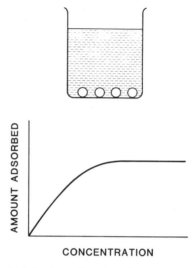

Figure 6.14. Isotherm for adsorption of catalytic ions.

Adsorption from solutions may be either cationic or anionic depending on the properties of the surface. In general, zeolites are strong cation exchangers, silica is a weak cationic adsorber, alumina adsorbs both cations and anions weakly, magnesia is a strong anion adsorber and carbon prefers to form charge-transfer complexes with electron donation but also weakly adsorbs cations.[157] With the exception of zeolites, that follow a different mechanism, the governing processes for ionic adsorption involve equilibrium reactions of the type

$$SOH^+ + C^+ \rightleftharpoons SOC^+ + H^+ \tag{6.6}$$

and

$$S(OH)^- + A^- \rightleftharpoons SA^- + (OH)^- \tag{6.7}$$

Equilibrium is determined by the concentrations of C^+ and A^- in the solution and by the pH. For cations, relative adsorption strength is largely a function of the polarizing power of the ions. A cation adsorber C, has affinities in the order $C^{4+} > C^{3+} > C^{2+} > C^+ \sim H^+$. Adsorption of anions depends on anionic polarizability and ionic charge, so that $SO_4^{2-} > I^- > Br^- > Cl^- > F^-$, etc. Ionic adsorption characteristics of catalytically important groups are given in Table 6.3.

TABLE 6.3. Adsorption of Catalytic Ions[a]

Group	Cation	Adsorption of element as		
		Cl^-	NO_3^-	SO_4^{2-}
1B	Cu^{2+}	Cationic < 4 M Anionic > 4 M	Cationic	Anionic
	Ag^+	Insoluble	Cationic (weak)	Anionic > 0.05 M
	Au^{3+}	Anionic > 0.1 M	Anionic (strong)	Anionic (weak)
VIII_3	Ni^{2+}	Cationic	Cationic	Cationic
	Pd^{2+}	Anionic > 0.1 M	Anionic	Anionic
	Pt^{4+}	Anionic > 0.1 M	Anionic	Anionic
VIII_2	CO^{2+}	Cationic < 6 M Anionic > 6 M	Cationic	Cationic
	Rh^{3+}	Anionic	Anionic	Cationic/anionic
	Ir^{4+}	Anionic > 0.1 M	Anionic	Anionic
VIII_1	Fe^{3+}	Cationic < 1 M Anionic > 1 M	Cationic	Anionic > 0.1 M
	Ru^{4+}	Anionic > 1 M	Anionic	Anionic
	Os^{4+}	Anionic > 0.1 M	Anionic	Anionic

[a] Reference 157.

Unfortunately, saturation amounts are generally small. With nickel solutions and alumina, for example, only loadings up to 2%–3% are possible. Multiple adsorptions with intermediate calcination give higher loadings, but this is time consuming. Other methods are generally preferred. With platinum and other expensive noble metals, however, amounts of less than 1% are often needed. Low loadings with high dispersion give satisfactory results. Supports are soaked in solutions of chloroplatinic acid, H_2PtCl_6, to yield desired levels of adsorbed $(PtCl_4)^{2-}$. Washing is not necessary, nor desirable, since it induces desorption. Drying and calcination are carried out as usual, with the chloroplatinic ion decomposing to platinum oxide or platinum.

Transport effects are encountered when using large particles. Adsorption of chloroplatinic acid is so rapid that diffusion of the solute into the pores controls the rate. Deposition takes place in an outer shell, as shown in Fig. 6.15.

In some cases, such profiles are desirable. For example, with fast reactions and external diffusion resistance, the reaction occurs only on the outside of the pellet. Platinum deposited deep inside the particle is wasted. Shell deposition is not satisfactory, however, in reactions that are not diffusion controlled, such as catalytic reforming. It can be avoided by adding hydrochloric acid to the solution. The HCl competes with chloroplatinic acid for adsorption sites, driving platinum deeper into the particle.

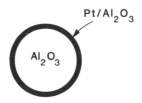

Figure 6.15. Shell adsorption of chloroplatinic acid.

This technique may be extended with strongly adsorbing acids such as oxalic, tartaric, and citric. These adsorb preferentially on the outside, forcing platinum ions deeper into the particles, as shown in Fig. 6.16.

Profile (a), for example, is useful in cases such as automobile exhaust oxidation. Platinum-poisoning lead contamination adsorbs in the outer platinum-free shell of alumina, which acts as a guard.[169] Applications for (b) and (c) include reactions with negative order and diffusionally limited, consecutive reactions.

6.4.3. Ion Exchange

Ion exchange in catalyst preparation is very similar to ionic adsorption but involves exchange of ions other than protons. Lower valency ions, such as Na^+, exchange with ions having higher charge, for example Ni^{2+}, according to the equilibrium

$$SNa^+ + Ni^{2+} \leftrightarrows SNi^{2+} + Na^+ \tag{6.8}$$

Ion exchange is useful in removing harmful agents and adding promoters. During washing with ammonium solutions, NH_4^+ ions are exchanged with impurities such as Na^+ to remove potential poisons. For controlled loadings of active components or promoters, the catalyst is soaked in excess solutions containing the exchange ions. Loading or extent of exchange is controlled with soaking time. Even multiple exchange is possible. Figure 6.17 shows

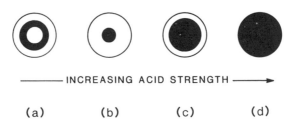

──────── INCREASING ACID STRENGTH ⟶

    (a)   (b)   (c)   (d)

Figure 6.16. Pellet concentration profiles with adsorbing acids.

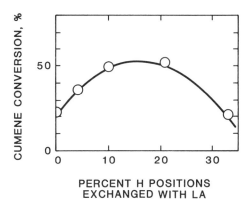

Figure 6.17. Ion exchange of NaY with H^+ and La^{3+} ions.[170]

the results of double exchange of NaY faujasite with $H^+(NH_4^+)$ and La^{3+} ions.[170]

Because of the large number of ion exchange possibilities, this method promises to be important for the modification of catalytic materials. For example, hydrous oxide ion exchangers of the type $Na_xTiO_2x(OH)$ are prepared with a wide concentration range of sodium, which is then exchanged with transition and noble metal ions.[171]

6.4.4. Impregnation

Also known as "incipient wetness," impregnation is the simplest and most direct method of deposition.[172] The object is to fill the pores with a solution of metal salt of sufficient concentration to give the correct loading. Figure 6.18 shows the sequence of steps. The support, usually in particle form, is heated or evacuated to remove pore moisture. This is not essential but speeds diffusion of the solute into the pores. Solution, in an amount just sufficient to fill the pores and wet the outside of the particles, is introduced. Although this may be calculated from measured pore volumes, it is sometimes more reliably determined with preliminary tests on aliquot samples.

Drying is necessary to crystallize the salt on the pore surface. If not performed properly, this step can result in irregular and uneven concentration distributions. For example, Fig. 6.19 demonstrates how the rate of drying affects pore and particle profiles.[173] If the rate is too slow, evaporation occurs at the miniscus, which retreats down the pore. Some salt deposition occurs but most of the solute merely concentrates deeper in the pore. When finally crystallized, the salt is located at the bottom of a pore

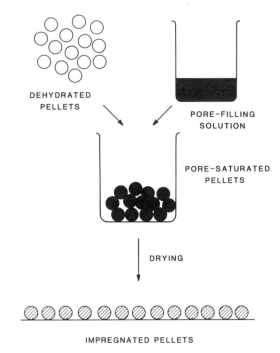

Figure 6.18. Steps in impregnation of the active component.

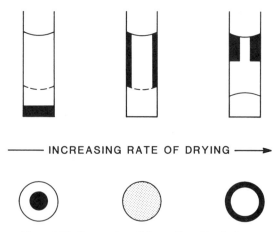

Figure 6.19. Pore and particle profiles after drying.

or at the particle center. When the drying rate is too fast, a temperature gradient occurs. Vaporization deep in the pore forces solution toward the outside, where most of the deposition takes place. The ideal situation is when crystallization is slow enough to form uniform deposits. However, since the support exists with a distribution of pore sizes, it is impossible to satisfy optimum conditions for each. Only experiment can establish the best procedures, but some nonuniformity must always be expected. When concentration profiles are desired for process reasons, these effects may be used to good advantage.[174,175]

Calcination is important in these circumstances. Crystallized salt redissolves when the dehydrated catalyst is exposed to moist environments and subsequent process drying may violate optimum conditions. Calcination converts the salt to an oxide or metal and essentially "freezes" the distribution. Other calcination effects, such as solid state reaction, also take place.

6.4.5. Activation

Activation is the final step in producing the deposited active component. None is necessary if the oxide itself is the active state. Conditioning by the process may be necessary, and this is examined in Chapter 8. If metals or sulfides are required, then reduction or sulfiding is necessary.

In reduction, the deposited oxide is converted to the metal by treatment with hydrogen, although other reducing agents such as CO or hydrazine are also used. The amount of metal produced depends on which oxidic compounds are present. For example, calcination of nickel salts deposited on alumina results in an increasing amount of $Ni[Al_2]O_4$ at higher temperatures. Since $Ni[Al_2]O_4$ is difficult to reduce, high concentrations influence the final metal content as shown in Fig. 6.20.

Partial reduction is common, leading to some uncertainty about the composition of the catalyst in the active system. Figure 6.13 shows nickel crystallites in contact with $Ni[Al_2]O_4$, which may play some role in the catalytic step. If $Ni[Al_2]O_4$ is not a factor, then better results may be possible by eliminating calcination and reducing the deposited hydroxide or salt directly.

The temperature of reduction is also important. Metal crystallite formation follows a sequence in which divalent nickel ions in the surface are first reduced:

$$Ni^{2+}-O^{2-} + 2H^+ \rightarrow Ni^0 + H_2O \qquad (6.9)$$

Protons are necessary and are produced through hydrogen dissociation by NiO itself. However, this requires some minimum temperatures, around

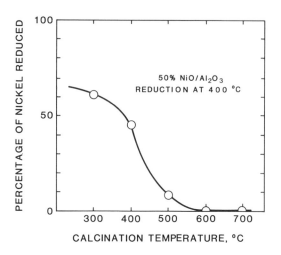

Figure 6.20. Effect of calcination on the reduction of coprecipitated $NiO-Al_2O_3$.[176]

300°C. Where lower reduction temperatures are desired, initial dissociation is facilitated by adding small amounts of CuO or Pt.[177]

The next step is nucleation of the nickel atoms into crystallites

$$Ni^0 + (n - 1)Ni^0 \rightarrow nNi^0 \qquad (6.10)$$

The rate of nucleation depends on the mobility of Ni^0 atoms, which is a function of both temperature and the nature of the substrate. Oxides that reduce with difficulty, such as $Ni[Al_2]O_4$, result in lower mobilities. The relative rates of reactions (6.9) and (6.10) determine the subsequent crystallite size and distribution. Fast reduction and slow nucleation give narrow distributions of small crystallites. Similar rates lead to broad distributions, and rapid nucleation to large crystallites. For a given substrate, temperature is the determining factor. Figure 6.21 shows the resulting crystallite size distribution for reduction of Ni/SiO_2 at 400°C and 500°C.[178] The lower the temperature, the better the dispersion.

Another factor is hydrogen purity. Reaction (6.9) is reversible. The presence of steam, even in small amounts, influences the rate and extent of reduction. Water is difficult to avoid even with high-purity hydrogen, since it is a product. When removal of reaction water is essential, the effect of the velocity of the hydrogen flow becomes significant, as shown in Fig. 6.22.[178]

The sensitivity of these parameters introduces grave doubts concerning control and reproducibility of reduction procedures. Very careful instructions must be followed for satisfactory results. Is it then advisable to carry

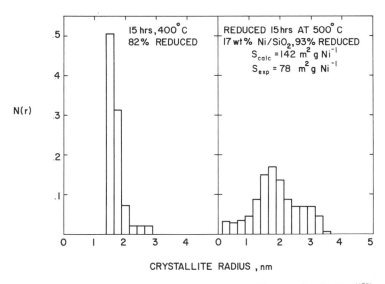

Figure 6.21. Effect of reduction temperature on crystallite size distribution.[178]

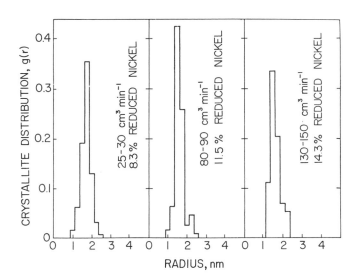

Figure 6.22. Effect of hydrogen velocity on the extent of reduction for 25% Ni/SiO_2 at 400°C.[178]

out reduction during preparation, where conditions are more uniform. Pre-reduction for commercial operations has many advantages. Nonuniformity due to impure hydrogen and poor flow control is avoided, start-up time is decreased, process hydrogen is not needed, and catalyst damage during reduction is eliminated. There are also benefits, even in laboratory use. If a catalyst charge is reduced in a large batch and used for separate testing and characterization, uniformity of the metal dispersion is more dependable. Unfortunately, high dispersions of metals are often pyrophoric and burn when exposed to air. They cannot be handled or shipped in the reduced state but must first be passivated. Passivation is accomplished in two ways. In the first method, the reduced catalyst charge is cooled and remaining hydrogen removed with a flow of inert gas. Oxygen or air is introduced into the inert stream at low concentration (less than 1%) slowly enough that the exothermic oxidation reaction does not cause hot bands. Oxidation is restricted to the first few layers of the metal surface, effectively protecting the bulk from further reaction. The catalyst can be handled safely. Only a small amount of surface reduction is necessary after loading in the reactor. Tests show that the original metal dispersions are retained.[178]

The second method is based on observations that the cause of pyrophoricity is not the reaction of oxygen with nickel but with adsorbed and occluded hydrogen atoms.[178] Residual hydrogen is removed by heating in an inert gas at temperatures 10–20°C above the reducing conditions. The catalyst may then be exposed directly to the atmosphere without burning. Passivation takes place through adsorption but without the painstaking steps used in the first method.

Another example of activation is found in hydrodesulfurization processes. Prepared as dispersed molybdena on alumina and promoted with cobalt or nickel, these catalysts are sulfided before use. In the plant, this is done either with a sulfur-containing feed or by pretreatment with CS_2 or H_2S. For laboratory operations, sulfiding with 10% H_2S/H_2 mixtures is sufficient. Care must be taken not to reduce in H_2 prior to sulfiding since the reduced state is less easily sulfided. Exact conditions, such as temperature, flow, or time, depend on the properties of the catalyst and the method of preparation, so that the sulfiding sequence is a factor in establishing activity.[179] Arguments for presulfiding before handling and shipment are also valid, and this practice is now becoming common in the industry.

6.5. EXTRACTION

Each method discussed so far uses some form of deposition to produce dispersion. It is also possible to generate porosity by extracting a component of the material. Examples are high-loading dispersed metals, porous metals, and reduced fused oxides.

6.5.1. High Loading Dispersed Metals

Coprecipitated oxide systems such as NiO–Al_2O_3 may be prepared over the complete composition range. High concentration, highly dispersed metal catalysts are produced by reducing the oxide structure. Compositions as high as 70–80 wt % Ni/Al_2O_3 with crystallite sizes of 2–3 nm are formed in commercial processes. Differences between low loading of nickel catalysts by adsorption or impregnation and high loading by coprecipitation are shown in Fig. 6.23. The intimate interaction between nickel and aluminum oxides in coprecipitation leads to the occlusion of small amounts of Al_2O_3 in the nickel crystallites upon reduction. Termed "paracrystallinity," defects so created are believed to impart increased activity and stability.[164] Another important example of this class is Cu–ZnO–Al_2O_3, used for low-temperature shift and methanol synthesis.

6.5.2. Porous Metals

There is a class of active porous metals known as Raney metals. Although nickel is the most common, preparations are possible with Fe, Co, and Cu. These metals form alloys with Al, which is easily leached out with caustic, leaving behind a porous metal with high surface area.[180] Raney nickel is prepared by melting nickel (mp 1230°C) and adding the required amounts of Al (mp 660°C). The alloy melt is thoroughly mixed and rapidly quenched in cold water. After crushing and sieving to 0.3–0.5 nm, 40 wt %

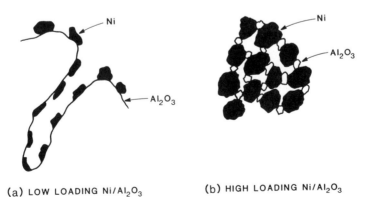

(a) LOW LOADING Ni/Al$_2$O$_3$ (b) HIGH LOADING Ni/Al$_2$O$_3$

Figure 6.23. (a) Low loading and (b) high loading in nickel catalysts.

caustic soda at 50°C is added slowly to give the reaction

$$Al + OH^- + 3H_2O \leftrightharpoons Al(OH)_4^- + \tfrac{3}{2}H_2 \qquad (6.11)$$

with evolution of hydrogen. The catalyst is washed free of caustic and stored under distilled water. These catalysts are used for a wide range of selective hydrogenation reactions using liquid phase slurry reactors.

6.5.3. Reduced Fused Oxides

The ammonia synthesis catalyst, Fe–Al_2O_3–K_2O, is the best example of reduced fused oxides.[2] Iron oxide, Fe_3O_4, and appropriate amounts of Al_2O_3 and K_2O are fused together at about 1500°C. When thoroughly mixed, the melt is cooled, crushed and sieved for use. Sizes range from 1 to 2 cm. Loaded directly as the oxides, the catalyst is reduced with feed gas, which removes the oxygen component of the Fe_3O_4 lattice, leaving behind a matrix of Fe crystallites with high porosity. Collapse of the structure at reaction temperatures (400–500°C) is prevented by the alumina, serving as a spacer. Promotion is achieved by the K_2O, which is believed to poison acid sites preventing the release of NH_3, to induce exposure of [111] planes of higher activity, and to donate electrons to the iron so that nitrogen chemisorption is aided.[14]

6.6. SPECIAL TYPES

Here we consider common types of catalysts not included in previous descriptions.

6.6.1. Mixed Oxides

In cases where intimate interaction is not necessary and where no synergistic effects exist, mechanical mixing of two oxides (or other components) may be all that is necessary.[181] This is accomplished by mulling with suitable mechanical devices. Early methanol catalysts, such as ZnO–$ZnCu_2O_4$, were prepared in this manner.

6.6.2. Cemented Oxides

When mechanical strength and thermal stability are paramount, as in steam reforming, this is accomplished by cementing components together with some type of ceramic cement, such as calcium aluminate.[182] Early steam reforming catalysts were of this type.

6.6.3. Metal Gauze

Platinum and silver gauze are used in ammonia and methanol oxidation to nitric oxide and formaldehyde, respectively. The gauze consists of fine wire mesh of 0.5 mm diameter, supported as layers within the reactor. Activation that "pits" the metal wire is necessary to enhance the active surface area.[183]

6.6.4. Honeycombed Monolith

The introduction of automobile exhaust catalysts in the United States and elsewhere has produced a major market for platinum-type oxidation and reduction systems.[184] An innovative consequence of this industry has been the development of ceramic honeycombed monoliths as catalyst supports. These structures contain long, parallel channels of less than 0.1 mm in diameter, with about 50 channels per square centimeter. The monolith is composed of cordierite ($2MgO \cdot 2Al_2O_3 \cdot 5SiO_2$) and is manufactured by extrusion.[185] A wash coat of stabilized alumina is administered prior to deposition of the active metal, either by adsorption or impregnation methods.

These structures have high mechanical strength, good thermal stability, and produce low pressure drop. Applications other than automobile exhaust clean-up will no doubt be forthcoming.

6.7. COMMERCIAL MANUFACTURE OF CATALYSTS

A current list of commercial catalyst manufacturers is given in Appendix 7. These companies account for almost two billion dollars worth of sales annually.[186] They contribute skills, expertise, and know-how that are unique and often the key to successful manufacturing. The business is highly competitive and secretive. Very little is available in the open literature concerning specific technology as practiced by these manufacturers. Nevertheless, it is important to consider their methods and problems, their expectations and limitations. If these factors are addressed early in catalyst development, then not only are impractical paths avoided but designers may find useful guidance from past experiences.

With the exception of particle formulations, catalyst manufacturing follows the same paths already considered in this chapter, except on a larger scale. In the laboratory, it is sufficient to make small batches—a hundred grams or a kilogram at the most. Even quantities this small often pose difficult quality control and manipulation problems. When the development moves to pilot unit testing or commercial demonstration stages, then tens or even hundreds of kilograms are required. Commercialization may demand plant capacities approaching hundreds of thousands of kilograms per year. Commercial catalysts must not only faithfully reproduce the performance of laboratory preparations but must also be uniform, consistent, and economical. A typical production line is shown in Fig. 6.24. Exact configurations vary from one product to another but follow similar routes.

Although catalyst manufacturing follows the same procedures as laboratory preparations, scale-up to marketable products introduces considerations not encountered in small-scale operations.[182]

6.7.1. Duplication and Scale-Up of Laboratory Recipes

During the course of normal catalyst development, a successful recipe is finalized and negotiations opened with a manufacturer to move to pilot testing and possible commercialization. Before serious talks proceed, the manufacturer will insist on testing the reproducibility of the recipe and attempt to make the catalyst from the laboratory instructions. This is to ensure that critical factors are understood and are not unknown artifacts of specific conditions. As strange as it may seem, cases are known where atmospheric dust in the laboratory was a key ingredient to nucleation of a gel and other environments did not work. This emphasizes the importance of keeping complete records during the research stages, not only for patent reasons but also as a source for tracing irregularities. Once the manufacturer satisfies himself that the recipe is accurate, then he can proceed with the question of scale-up.

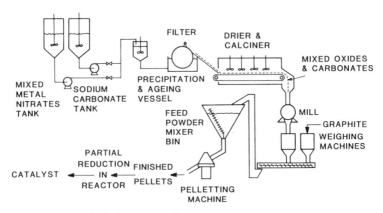

Figure 6.24. Typical production line for catalyst manufacturing.

Each step in the recipe must be translated into large-scale operation. First, each must be technically feasible, that is, engineering counterparts must exist for laboratory procedures. Equipment is usually no problem,[35,160] but maintaining the controlling conditions may be. Second, scale-up must be economical. The most important factor here is capital investment to acquire facilities for what may be large quantities. Purity levels of reagents is another issue which requires attention. Large quantities of very pure feedstocks may be expensive or even unattainable.

Equipment is usually available or adaptable for most steps. Table 6.4 list common unit operations found in catalyst manufacturing. Many are material treating methods used by the solid or mineral processing industries. Relative difficulties in scale-up have already been touched upon; for example, drying is usually easier, precipitation harder.

6.7.2. Continuous Unit Operations

As in all chemical processing, continuous operation is much more satisfactory for large throughputs. Operating costs are less and product quality control better. Whenever possible, the catalyst manufacturer attempts to modify preparational procedures to accommodate continuous operations. This is fairly straightforward for filtering, drying, calcination,

TABLE 6.4. Unit Operations in Catalyst Manufacturing

Preparation step	Unit equipment
Precipitation	Tanks, batch and semibatch
	Mixing tanks
	Jets
Filtering	Leaf
	Rotary
	Centrifugal
Drying	Trays
	Forced air
	Moving belts
	Rotary kilns
	Fluidized beds
	Spray driers
Calcination	Forced air furnace
	Rotary kiln
	Fluidized beds
Formulation	Pelleting
	Extrusion
	Spheroidization

and formulation, but is difficult for precipitation and deposition. These are usually handled by batch or semibatch processes, with intermediate storage to ensure continuous downstream processing. Continuous adsorption or impregnation is practiced with belts or baskets moving through tanks or with spraying.

6.7.3. Energy Conservation and Environmental Control

Washing, drying, and calcination in the laboratory do not pose serious energy or pollution problems. With large-scale processing, heating steps must be adapted to minimize energy consumption. This is possible with good engineering design and control. Water and volatile effluents from washing and heating must be cleaned before discarding or reuse. In severe cases, expensive processing may be necessary.If at all possible, manufacturers try to minimize these problems through modification of the precipitation steps.

6.7.4. Multiproduct Flexibility

Some products, such as catalysts for cracking, hydrotreating, and reforming, are consumed in such large quantities that plants may be dedicated to their production. Most catalysts, however, do no justify special facilities, especially in pilot testing. Manufacturers prefer to use existing equipment for producing different catalysts. An available facility may not be the most optimum, but its use could result in substantial savings, even if modifications of the recipe are required.

6.7.5. Confidentiality, Secrecy, and Know-How Protection

Catalyst manufacturers face special problems with confidentiality, secrecy, and know-how protection. Because of the nature of their business, they enter into confidentiality agreements with many customers. Care must be taken to ensure that proprietary information does not pass outside existing and past agreements. Secrecy is very strict, sometimes even within a single organization, where different groups may be collaborating with competing customers.

Solutions to problems in scale-up become part of the manufacturer's assets. Such solutions are keys to commercial successes and constitute competitive advantages. For this reason, technology within a plant is carefully guarded and protected.

6.7.6. Competition and Brief Pay-Out Periods

Competition within the industry is very intense. When a new catalyst emerges, there are usually similar developments not far behind. Patent coverage is important but becomes confused very quickly. New areas bring a flood of patent applications, which are only clarified after a long period of litigation. In the meantime, the frontrunners attempt to capitalize on their advantage. Establishing a strong market position early is essential, since a new venture has about three to four years before competitive pressure is felt.

6.8. CATALYST FORMING

The only processing step not yet considered is catalyst forming, since this is only practiced in commercial manufacturing. Laboratory usage does not normally include particle formation, powder samples being sufficient for most purposes. Indeed, formulation techniques are very difficult to duplicate on a small scale and it is better to leave these to the experts. We consider here the principal factors in the production of common forms of particles.

6.8.1. Pellets

Pellet or pill formation is illustrated in Fig. 6.25. The process is essentially powder compression in a pelleting press. Powder is poured into a

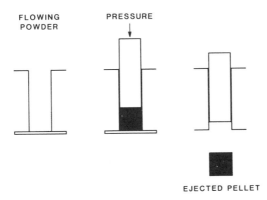

FLOWING POWDER   PRESSURE

EJECTED PELLET

Figure 6.25. Production of pellets.

cylindrical cavity as shown and a piston applies prescribed pressures. Values from 10^2 to 4×10^3 atmospheres are common, depending upon the compressibility properties of the powder. Grains must be deformable since boundaries flow together to produce the finished pellet. Factors such as the ultimate tensile strength of the materials, mesoporosity of the grains, and moisture content are important. Some materials, such as kieselguhr, pellet easily, others require assistance from the addition of small amounts of plasticizers and lubricants. Graphite, talc, and stearic acid are used for this purpose.[35]

The pellet is ejected after compression. It is very uniform in shape and dimensions, has high mechanical strength and moderate porosity. Fibers are sometimes added, for example, polymers, that are burned off to increase macroporosity, and metals, to improve thermal conductivity in or out of the pellet.

Pelletization is the most expensive method. For continuous production, complicated disk-type rotating cylinders with staggered pistons are used. Complexity of these movements, stress on metal parts due to high pressures, and presence of abrasive powders all contribute to high equipment and maintenance costs. For a given throughput, larger pellets are generally cheaper to make.

6.8.2. Extrudates

Figure 6.26 shows the formation of extrudates. A slurry of the catalyst powder is fed from a hopper into the screw drive. Peptizing agents, such as nitric acid, may be added to deagglomerate the primary particles by lowering the zeta potential and improve the extrusion process.[188] The screw forces the slurry through holes in the end plate. Usually circular in cross section, these holes can also be made in the shape of lobes or stars. As the ribbon of slurry emerges from the hole, it begins to dry and harden sufficiently to maintain its shape. The ribbon is either cut into prescribed

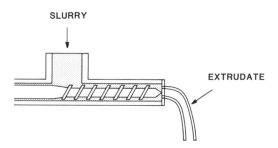

Figure 6.26. Production of extrudates.

lengths by a knife rotating outside the end plate or allowed to break up as it falls onto a moving belt on its way to the drier.

Extrudates are irregular in shape and length, have higher porosities and lower strengths than pellets, but are less expensive to make. Many commercial products are now available as extrudates.

6.8.3. Spheres

Hydrogels that age rapidly, such as alumina, silica, and silica-alumina, are formed into spheres by a technique similar to the historical method for making musket balls. The hydrogel is forced through holes in a plate at the top of a column containing oil in which the gel is immiscible, as shown in Fig. 6.27. Assuming a spherical shape, the drop falls slowly through the oil, hardening as it ages. This process is accelerated by increasing the pH down the column with flowing ammonia up the tower. At the bottom, the spheres are separated, dried, calcined, and sieved. Very uniform spheres are used in moving and ebullating beds.[189]

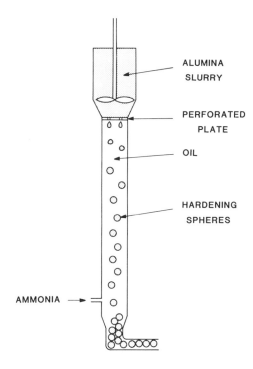

Figure 6.27. Production of spheres using the column method.

Another method uses an inclined disk, as shown in Fig. 6.28. Moist powder is applied to the pan. A small particle falls into the pan and grows as it rotates. In a different adaptation, a drop of hydrogel emerges from the center as the plane rotates. Moving under centripetal force, the drop hardens, becomes spherical, and finally falls off the plate.

6.8.4. Flakes and Pastilles

A very important catalytic process is hydrogenation of fats, fatty acids, and vegetable oils used in foodstuffs.[190] Reduced nickel catalysts are used for this purpose. They usually contain about 60% Ni on a support and are suspended in a protective hard fat, such as saturated glyceride, tallow, or vegetable oil. The particles are flake- or drop-like and are used in liquid phase hydrogenation after the protective fat dissolves in the reagent. Since the fat becomes part of the product, catalysts used in food production must conform to health and religious standards.

6.8.5. Granules

Materials manufactured from fused oxides are not amenable to pelleting or extrusion. The best approach is to use them as specific mesh ranges of crushed granules. Sizes available are 8–14, 6–8, 4–10, 4–6, and 2–4 mesh. Particles are irregular in shape, and this is some advantage in bed porosity. However, if the range is too wide, small particles pack into interstices

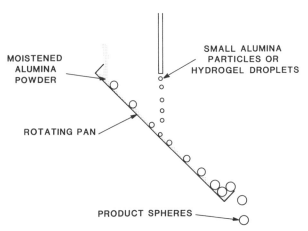

Figure 6.28. Production of spheres using the rotating dish method.

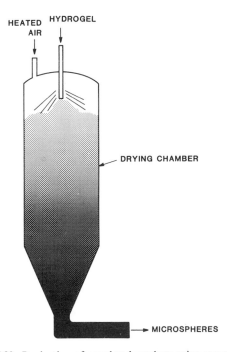

Figure 6.29. Production of powdered catalysts using spray drying.

between larger ones, reducing porosity and increasing pressure drop. Better results are found with uniform sizes.

6.8.6. Powders

Fluidized bed and slurry reactors utilize catalysts directly in the form of powders (50–500 μm) and microspheres (up to 1.0 mm). These are best produced by spray-drying, as shown in Fig. 6.29. Hydrogel is sprayed through nozzles into a heated zone. Drying and calcination occurs rapidly as the small drops fall.

6.9. MECHANICAL STRENGTH

Formulation of particles is the final step in manufacturing and establishes the mechanical strength of the catalyst. This is an important engineering parameter and manufacturing methods should satisfy process specifications. Catalyst particles are designed to withstand the following types of mechanical stress:

6.9.1. Abrasion During Transit

Abrasion is loss of material due to contact of particles with container walls. Commercial catalysts are shipped to customers in metal, plastic, or fiber drums containing up to 250 kg. Polyethylene liners or bags are often used. Inert blanketing is usually not necessary. Drums are transported by train or truck, handled by fork-lift in and out of storage, and lifted to the top of reactors with cranes. Care must be taken to avoid excessive shaking, rolling, dropping, or other mishandling during these operations. Drums should not be exposed to inclement weather conditions. It is advisable to use special rigs for effective movement of the drums, as recommended by the manufacturer.

6.9.2. Impact During Reactor Loading

Reactors are large vessels, often several meters in diameter. As much as 5×10^4 kg of catalyst is loaded at one time. Access is through man-holes at the top, leaving considerable distances up to 10 m to be filled. Catalyst particles must not be poured directly into the vessel, since the fall could shatter them and result in a layer of fragments at the bottom. Various devices such as buckets and sleeves, as shown in Fig. 6.30, are recommended. Operators entering the vessel to smooth the layers should be careful not to exert too much weight on small areas of the bed and should protect themselves against dust and toxic hazards.

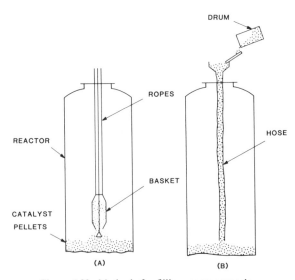

Figure 6.30. Methods for filling reactor vessels.

6.9.3. Internal Stresses from Phase Changes During Such Processes as Activation and Regeneration

Mechanical properties of the manufactured catalysts do not necessarily match those after activation and regeneration. Strengths should not change too much as the particles weaken due to phase changes during start-up and use. It is not uncommon for packed beds to shrink considerably when operation commences. Problems are avoided by following loading and operating instructions exactly.

6.9.4. External Stresses Caused by Fluid Flow, Pressure Drop, Catalyst Bed Weight, and Thermal Cycling

The range of strengths that are required are usually found from experience. A given reactor design will address this problem, but care should be taken not to violate specifications in selecting a fresh load of catalyst.

6.9.5. Attrition (Particle–Particle Collision Losses)

This is only a problem with moving or fluidized operations. Catalyst particle strength must match specific requirements. Attrition loss is usually determined by some standard test, described in Chapter 7.

6.10. LEGAL RESPONSIBILITIES

Catalyst design, preparation, manufacture, and use is a sensitive business. Very small changes in composition or steps in the preparation can result in advantages in processing. There is a very small time scale between invention and implementation. It is understandable that organizations engaged in these enterprises are very protective of proprietary information. Workers in the field must appreciate and honor this. Not only should every effort be made to maintain security, but information from others should not be accepted. Discoveries and developments have been jeopardized in the past because researchers unwillingly or illegally had knowledge of data from competitors working in the same area. Always exercise secrecy agreements and honor them.

Commercial catalysts are available under the following conditions described below.

6.10.1. Proprietary

The process is licensed from the vendor and includes the catalyst. The licensing organization selects the original catalyst and subsequent charges but guarantees their performance. In some cases, the catalyst is only "rented" and is returned when used. All operations are supervised to some degree by representatives of the vendor. Very little information is revealed and the user is restricted not to make any tests or analyses other than those prescribed for operations.

6.10.2. Partly Restricted

The process licensor specifies the initial charge of catalyst, with or without restriction on information, but the user may select subsequent loadings from competing producers.

6.10.3. Open Market

The user may select from a wide range of catalysts available from many manufacturers. Legal restrictions vary from one case to another.

6.10.4. Developmental

Manufacturers often have semicommercial batches of catalysts available for testing. Negotiation and secrecy agreements are usually required.

6.11. SOME FINAL THOUGHTS

There is no doubt that catalyst preparation is the secret to achieving the desired activity, selectivity, and lifetime. Catalyst manufacturing, in turn, is the key to commercial utilization and profit. Raney's modern alchemist now has many sophisticated tools to aid him, yet diligence, perseverance, and perhaps luck, are still necessary. Let those who tread this path take warning and ensure that all parts of the puzzle fit and are properly recorded.

7

CATALYST CHARACTERIZATION
How They Are Tested

7.1. PRELIMINARY REMARKS

Characterization of the catalyst is necessary at every stage of development. Critical parameters are measured not only to check the effectiveness of each operation but also to provide specifications for future products. Preparational procedures can be properly assessed and corrections made only if guidelines to the properties are available. Product quality in commercial production is facilitated with appropriate testing, and the first step in any diagnosis of process problems is to measure changes that have occurred. It is vital that accepted techniques for determining these characteristics be established as an adjunct to research design, preparation, testing, and manufacture.

In this chapter, we examine current methods for measuring those properties of the catalyst relating to plant performance. These are classified as bulk, particle, and surface properties, including such factors as composition, structure, mechanical properties, surface area, dispersion, and acidity.[191] Although these measurements also enter into research, specialized techniques in scientific studies are not included, nor do we treat methods that investigate adsorbates, unless they are part of adsorbent characterization.

With this approach, we describe applicability of each method, including capabilities and limitations. Most workers in catalyst development are not necessarily expert in analytical procedures. The techniques described are usually available within the facilities of most companies or from outside testing laboratories. For this reason, details of instrumentation and methodology have been minimized in order to concentrate on what should be expected and requested, and to present guidelines for the selection of appropriate tests.

Historically, organizations in catalyst development and use have evolved their own analytical procedures and equipment. Although similar in many respects, sufficient differences emerged to make precise comparisons meaningless. For example, measurements of particle crushing strength were reported not only in a wide range of unit systems but also in unit dimensions. Test results were dependent on the apparatus used, so that relative assessment was pointless. Recognizing that characterization is one area where cooperation is possible and desirable, the industry has taken measures to standardize testing procedures and equipment. In the United States, a committee of the American Society for Testing of Materials (ASTM) was established in 1974 to develop accepted methods.[192,193] Similar but less organized efforts started in Europe,[194] the Soviet Union,[195] and Japan.[196] The ATM Committee D-32 is now issuing detailed descriptions and more will be forthcoming. Whenever possible, this chapter conforms to recommendations from the committee's deliberations.

7.2. BULK PROPERTIES

The most important bulk properties for either powders or particles are (1) composition, and (2) phase structure.

7.2.1. Composition

Qualitative and quantitative identification of elemental components of the catalyst is a basic necessity. This includes not only the principal and impurity components added during preparation but also contamination deposited during use. These deposits include: (1) rust and other debris, (2) poisons from the feed, such as S, As, Pb, and Cl, (3) secondary contaminations in the feed such as Ni, Fe, V, Ca, and Mg, and (4) carbon deposited as coke during fouling. In the case of particles, determination of the composition profile is useful in diagnosis. The field of analytical chemistry has developed a vast array of methods. Physical and chemical, destructive and nondestructive, "wet" and instrumental, all are legitimate possibilities for application in catalysis. Those receiving the most attention are as follows.

7.2.1.1. Solution Methods

Each element is susceptible to some type of "wet" test, a fact familiar to every student of quantitative analysis. Usually the element of interest is dissolved in some form giving a characteristic color, with quantitization using photometric absorption. For example, standard procedures recommended for cobalt measurement in hydrodesulfurization catalysts are as follows[193]:

> Decompose the sample by heating with H_2SO_4, dilute, and select aliquots containing 10–30 mg of cobalt. Add each of these to measured volumes of potassium ferricyanide, ammonium citrate, ammonia, and ether. After cobalt has complexed with the ferricyanide, back-titrate with standard cobalt solution, using potentiometric titration. From the original concentrations (not given here) calculate cobalt concentration in the sample. No other constituents interfere. Alternatively, the cobalt complex concentration can be measured with photometric absorption.

This procedure is typical of those available for important catalytic components. Standard methods exist, for example, for cobalt, nickel, molybdenum, and platinum.

7.2.1.2. Spectroscopic Methods

Beginning with atomic emission spectroscopy, there is an arsenal of such methods available. Perhaps the most versatile and often-used technique is x-ray fluoresence, in which a sample is bombarded with hard x-ray photons. Secondary x-rays are ejected with wavelengths characteristic of the atom. Even small quantities of elements are detectable and measurable. Complete elemental analyses are available, rapidly and accurately, with small amounts of sample.[197]

An allied method is electron probe analysis.[198] Electrons impinge the sample to emit secondary electrons or photons characteristic of the element. Scanning is possible with small area probes, so that profiles across particle diameters can be measured. Figure 7.1 shows typical results.

Other techniques include atomic absorption spectroscopy, inductively coupled plasma spectroscopy, and analytical electron microscopy.[200] With techniques such as these available, catalyst developers have no difficulty obtaining accurate elemental analyses of their samples.

7.2.2. Phase Structure

Phase identifications is a much more difficult task. Catalysts contain many components, each of which may exist in several different structures. Variations in composition complicate interference and sensitivity factors. Nevertheless, considerable success is achieved in practice, and the knowledge of structure is usually extensive. The most successful approach is to measure either a diffraction or temperature programmed pattern and identify

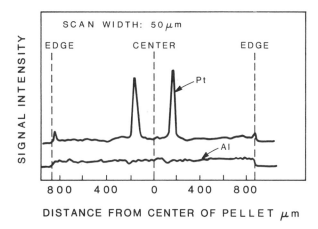

SCAN WIDTH: 50 μm

DISTANCE FROM CENTER OF PELLET μm

Figure 7.1. Profile scans with an electron probe analyzer.

components by matching "fingerprints" from pure compounds. Common methods are the following.

7.2.2.1. Diffraction Methods

X-ray diffraction is well developed and usually gives satisfactory results.[197,201] Monochromatic x-rays are reflected from the sample with diffraction lines produced from the repetitive dimension of crystal planes. Each crystal type gives a characteristic pattern, so that the position of lines is a clue to the presence of a particular compound. Figure 7.2 shows typical results for alumina catalysts.

There are three complications: (1) A minimum amount of material, depending on atomic weight, is necessary for detection; usually, 1–5 wt % is required. (2) Diffraction lines broaden as crystallite size decreases; discrimination is difficult with crystallites less than 5 nm in diameter. (3) Lines from different components often occur in similar positions, or overlap and interfere with each other. Precise assignment becomes questionable. However, modern counting electronics and computer interpretation have increased accuracy to a point where many of these disadvantages are overcome.

Electron diffraction is also possible during electron microscopy.[202,203,204] Resolution is very high and individual crystallites may be characterized. Other attachments allow scanning electron probe and x-ray fluoresescence analysis over the same region. Neutron diffraction has also been used in some applications.[205]

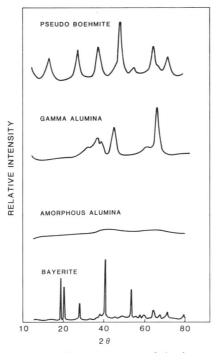

Figure 7.2. X-ray diffraction patterns of alumina catalysts.

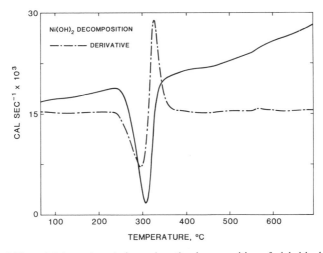

Figure 7.3. Differential thermal analysis results—the decomposition of nickel hydroxide.[162]

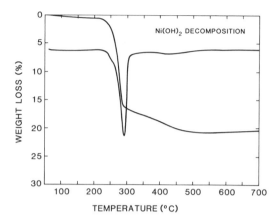

Figure 7.4. Thermal gravimetric analysis results—the decomposition of nickel hydroxide.[162]

7.2.2.2. Temperature Programmed Methods

Differential thermal analysis (DTA) and thermal gravimetric analysis (TGA) are the most useful. The first measures energy changes as the sample is scanned through phase changes[206,207]; the second records weight loss or gain.[206] Matching with standards is necessary for identification. Figures 7.3 and 7.4 show catalytic examples for each technique.

Additional information is obtained by using reactive atmospheres, usually hydrogen but also oxygen and H_2S. With hydrogen the technique, called temperature programmed reduction (TPR),[208,209] gives information on the reducibility of oxides. Catalysts that are easier to reduce, i.e., in which oxides are bound less strongly to the support, show reduction "peaks" at lower temperatures.

7.3. PARTICLE PROPERTIES

These are properties of the particle only, but include many important parameters. Here we shall consider (1) densities, (2) particle size, (3) mechanical properties, (4) surface area, (5) pore size distribution, and (6) diffusivity.

7.3.1. Densities

The question of defining and measuring densities may appear trivial. Density is mass per unit volume. However, the questions of which volume to use, what to call it, and how to determine it have generated much confusion in catalysis. Here we adopt the procedures of the ASTM Committee.[193] There are four major definitions of density: theoretical, skeletal, particle, and packing.

7.3.1.1. Theoretical Density

This is defined as the ratio of the mass of a collection of discrete pieces of solid to the sum of the volumes of each piece, if the solid has an ideal regular arrangement at the atomic level. Theoretical volumes are determined from x-ray diffraction unit cell measurements, so this density is also known as the x-ray or unit cell density. Ideal volumes are of little use in catalysis, and this term has hardly any applicability.

7.3.1.2. Skeletal Density

In this density, volume is defined as the sum of the volume of the solid material and any closed pores within the solid. These pores cannot be penetrated by any fluid and become part of the powder volume. A mass of catalyst is placed in a flask of known volume, and the amount of helium needed to fill the flask measured, giving the powder volume by difference. Care should be taken to dehydrate all pores thoroughly. Because helium is used as the displacing fluid, this density is sometimes called the helium density. See Table 7.1 for an example of typical values.

7.3.1.3. Particle Density

Here the volume is the sum of the solid, closed pores, and accessible pores within the particle. It is essentially the volume of the particle, but should not be found by measuring dimensions. A displacement pychometer is used but with a fluid that does not penetrate the interior pores of the pellet. One approach is to fill these pores with the fluid prior to displacement, for example, with methanol. A more satisfactory method is to use mercury,

TABLE 7.1. Example of Densities 7 wt% NiO/Al_2O_3

Density	Value
Theoretical, d_t	3.89 g cm^{-3} (crystal)
Skeletal, d_s	2.39 g cm^{-3} (solid + closed pores)
Particle, d_p	1.22 g cm^{-3} (pellet)
Packing, d_b	0.732 g cm^{-3} (bed)
	$\varepsilon = 0.40$
	$\theta = 0.49$

which does not penetrate pores smaller than 1.2×10^3 nm at atmospheric pressure. For this reason particle density is also called mercury density. There is a relationship between particle and skeletal densities, d_p and d_s, and porosity, θ, given by

$$\theta = (1 - d_p/d_s) \qquad (7.1)$$

7.3.1.4. Packing Density

Also called bulk or bed density, the volume in this case includes the void space between particles. This general definition poses some problems since measured volumes differ, for example, for various vessels, packing methods, and agitation. This is overcome by specifying dimensions of the container used and the method of packing. From a practical viewpoint, packing methods should be the same as those used in the reactor filling. Packing density, d_b, and particle density are related through the void fraction, ε; thus we have

$$\varepsilon = (1 - d_b/d_p) \qquad (7.2)$$

Values in the example in Table 7.1 show that considerable errors result if these densities are confused. A good practice is to always specify the volume included, as in Table 7.1.

7.3.2. Particle Size

Measurement of the particle size of macroscopic pellets, extrudates, and spheres presents no problems. Dimensions can be determined directly or sample material sieved for irregular particles. Smaller distributions in powders, such as cracking catalysts, require specialized procedures: optical and electrical imaging, light scattering, light shadowing, elutriation, sedimentation, electrical resistance, impaction, and nozzle pressure drop.[210,211,213,214] For cracking catalysts and other powders in the range 0.6–160 μm, the preferred approach is to use an electronic particle size analyzer. An electric current path of small dimensions is modulated by individual particle passage through apertures, producing individual pulses proportional to the particle volume. The sample is suspended in an electrolyte, dispersed ultrasonically, and forced through restricting apertures. Distributions are found from the number of particle passing through progressively smaller openings.[211] Table 7.2 gives typical results.

TABLE 7.2. Typical Results for a
Particle Size Distribution Silica–
Alumina Cracking Catalyst

Size range (μm)	Distribution (wt%)
0–20	3
20–45	35
45–60	25
60–90	25
90+	12

7.3.3. Mechanical Properties

Mechanical stresses experienced by catalyst particles during handling and use are considerable. The nature of these stresses is discussed in Chapter Six. Properties that relate to stress resistance are crushing strength, attrition loss, and loss on ignition.

7.3.3.1. Crushing Strength

The basis for evaluating crushing strength is the Huit equation[214]

$$F/D_p^2 = C \qquad (7.3)$$

where F is the force necessary to crush a particle of diameter D_p and C is a constant which depends upon (1) the ultimate tensile strength of the material, (2) the plasticity of the solid powder, (3) the grain density, (4) porosity, (5) pelleting pressure, (6) binders, and (7) devices used in testing. With such an array of variables, it is not surprising that existing correlations and techniques are empirical, with no scientific foundation. However, although only relative, these tests do index performance of particular formulations. Most commonly found are variations of axial and radial crushing tests, shown in Fig. 7.5. A statistical number of particles are crushed, either along the axial direction for cylinders or the radial direction for extrudates, and an average taken. Values range from 10 to 100 kg cm^{-2} for axial and 1–10 kg cm^{-1} for radial tests. The process designer knows from experience that particles with certain specifications give satisfactory results. Although initially based on performance in similar reactors, these specifications are continually updated.

A much more satisfactory procedure is to measure bulk crushing strength, which more closely approximates packed beds and is applicable

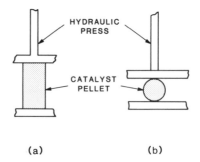

(a)   (b)

Figure 7.5. (a) Axial and (b) radial crushing test for catalyst pellets

to all shapes, pellets, extrudates, and spheres.[193] The recommended apparatus is shown in Fig. 7.6. A cylindrical sample holder 5 cm in diameter and 5 cm deep is filled with about 100 cm^3 of catalyst particles. The catalyst is dehydrated by heating in air at 200°C for three hours. A constant pressure is applied, from 3 to 30 atm for one minute. After pressure application, the bed is removed and sieved for fines, using the following U.S. Standard Series mesh for each catalyst size: 0.8 mm—40, 1.6 mm—20, 2.5 mm—12, and 6.5 mm—4. The weight percent fines below the minimum size are recorded. The experiment is repeated at different pressures to generate data of the type in Fig. 7.7. The bulk crushing strength is taken as the pressure necessary to produce one weight percent fines. ASTM tests showed excellent reproducibility for results from one organization to another.[193]

7.3.3.2. Loss on Attrition

This test simulates actual use by tumbling 100 g of catalyst in a drum at 56 rpm, as shown in Fig. 7.8. Attrited fines less than 20 mesh in size fall

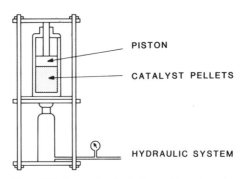

PISTON

CATALYST PELLETS

HYDRAULIC SYSTEM

Figure 7.6. Apparatus for bulk crushing strength.

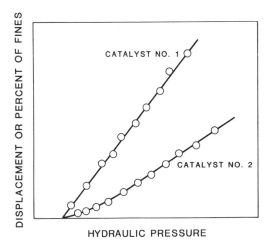

Figure 7.7. Results of bulk crushing strength tests.

through the screen and are collected. Losses of less than one weight percent are acceptable.

7.3.3.3. Loss on Ignition

Porous particles contain up to 20% moisture adsorbed in pores and on the surface. Calculations that use the weight of the catalyst should always correct for this. Exact moisture content is determined by heating a pre-weighed sample in a crucible at temperatures above 1000°C. All volatile material, adsorbed moisture, and any constitutional water is removed. The sample is cooled in a dessicator and weighed.

Figure 7.8. Apparatus for measuring loss on attrition.

7.3.4. Surface Area

The term "texture" refers to the general pore structure of particles and includes surface area, pore size distribution, and shape. Total surface area, S_g m^2 g^{-1}, is possibly the most important particle parameter specified without regard to the type of surface. No attempt is made to distinguish one component from another.

Measurement of surface area involves the principles of physical adsorption, which differ from those of chemical adsorption, as shown in Table 7.3.[23]

Physical adsorption is equilibrium coverage similar to surface liquefaction. Produced by van der Waals forces originating in surface atoms, interactions with the surface are similar to those between molecules and are approximately the same for all materials. Coverage proceeds first with adsorption on surface atoms but is quickly followed by the generation of additional layers even before a complete monolayer forms. Since the process is exothermic and at equilibrium, the amount decreases as temperature increases. Easily measurable quantities are found close to the normal boiling point of the adsorbate. Equilibrium isotherms follow the shape shown in Fig. 7.9, where the volume adsorbed is plotted against p/p_0 (p is the pressure, p_0 the saturation pressure at measurement temperature). Three regions are included. At low pressures monolayer formation follows the Langmuir equation

$$\frac{V}{V_M} = \frac{Kp/p_0}{(1 + Kp/p_0)} \tag{7.4}$$

TABLE 7.3. Differences between Physical and Chemisorption

Property	Physical adsorption	Chemisorption
1. Forces responsible	Physical van der Waals electrostatic forces	Chemical bonds, ionic, covalent
2. Heat of adsorption (exothermic)	Low (<10 kcal mole^{-1}) similar to liquefaction	Moderate to high (10–50 kcal mole^{-1}) similar to reactions
3. Activation energy	None	Low (<15 kcal mole^{-1}) to moderate (25 kcal mole^{-1})
4. Specificity	None	Specific adsorbate-adsorbent interactions
5. Reversibility	Complete, rapid	Slow
6. Extent	Multilayers	Monolayer saturation

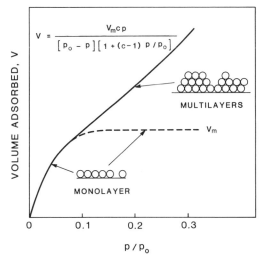

Figure 7.9. Typical isotherm for physical adsorption.

with V_M the monolayer volume and K a constant. However, above approximately $p/p_0 = 0.1$, multilayer condensation is extensive. A theoretical model for this was first derived by Brunauer, Emmett, and Teller, giving the now famous BET equation[215]:

$$\frac{V}{V_M} = \frac{cp}{[p_0 - p][1 + (c - 1)p/p_0]} \qquad (7.5)$$

The parameter c includes the heats of adsorption and liquefaction and is fairly constant for a given class of materials (e.g., oxides and metals), with values below 100.

Equation (7.5) is valid up to $p/p_0 = 0.3$. Above this point, liquid condensation begins in the smallest micropores and progresses through mesopores as p/p_0 approaches unity. This feature is discussed in the section on pore size distributions.

Finding V_M by fitting equation (7.4) is impractical, since it is valid in such a small region. Early workers attempted to estimate the monolayer graphically, with resulting imprecision and irreproducibility. The BET equation was a breakthrough in catalytic science by providing an accurate and reliable means of finding V_M. Rearranging variables gives

$$\frac{p}{V(p_0 - p)} = \frac{1}{V_M c} + \frac{(c - 1)}{V_M c}(p/p_0) \qquad (7.6)$$

leading to BET plots shown in Fig. 7.10. Very linear results are found and
the slope, S, and intercept, I, are easily measured. The example in Fig. 7.10
gives

$$S = 13.85 \times 10^{-3} \, \text{cm}^{-3}$$

$$I = 0.15 \times 10^{-3} \, \text{cm}^{-3} \qquad (7.7)$$

$$V_M = 1/(S + I) = 71 \, \text{cm}^3 \, (\text{STP})$$

and, for 0.83 g of sample,

$$S_g = \frac{(71)(6.02 \times 10^{23})(16.2 \times 10^{-20})}{(22{,}400)(0.83)} = 373 \, \text{m}^2 \, \text{g}^{-1}$$

Data in Fig. 7.10 were taken with nitrogen, whose cross-sectional area when
packed on the surface is $0.162 \, \text{nm}^2$. Nitrogen is a readily available and
preferred adsorbate. Isotherms are usually measured volumetrically, using
an apparatus of the type shown in Fig. 7.11.

In this apparatus, four samples are loaded at one time. This increases
productivity, since the most time-consuming step is dehydration. If preferred, a separate apparatus may be used for degassing any number of
samples simultaneously. After dehydration at 300°C (or lower for tem-
perature-sensitive samples) in vacuum for extended periods, three samples
are isolated while the fourth is measured. The flask is cooled in a constant
level of liquid nitrogen, the valve closed, and helium admitted to the known
volume at some pressure. When the valve is opened, helium fills the "dead"

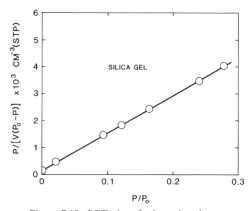

Figure 7.10. BET plot of adsorption data.

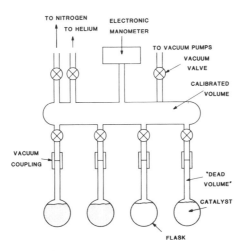

Figure 7.11. Apparatus for adsorption measurements.

volume and the pressure drops. This gives the dead volume, which should be as low as possible. The flask is evacuated and the procedure repeated with nitrogen. Some of the nitrogen is adsorbed so that the new pressure is lower than predicted, thus giving the volume of nitrogen absorbed by difference. The process is continued by incremental adsorption on the surface until the complete isotherm is obtained. Typical values for S_g are given in Table 7.4.

In practice, it is inconvenient to measure the number of points shown in Fig. 7.10. Usually a minimum number are taken (2 or 3), each requiring about thirty minutes. Indeed, for routine determinations, c is often known with enough confidence to justify one point. Even if c is not available, equation (7.7) shows that ignoring the intercept introduces only a small

TABLE 7.4. Typical Surface Areas for Catalysts

Catalyst	Use	S_g (m^2 g^{-1})
REHY zeolite	Cracking	1000
Activated carbon	Support	500–1000
SiO_2–Al_2O_3	Cracking	200–500
$CoMo/Al_2O_3$	Hydrotreating	200–300
Ni/Al_2O_2	Hydrogenation	250
Fe–Al_2O_2–K_2O	Ammonia synthesis	10
V_2O_5	Partial oxidation	1
Pt gauze	Ammonia oxidation	0.01

error, and with $c \sim c - 1$ we have

$$V_M = V(1 - p/p_0) \tag{7.8}$$

as the basis of the "one point" method in many automated instruments. Other approximations used are[216]

$$V_M = 3.5 V_{p/p_0 = 0.2} \tag{7.9}$$

or

$$V_M = V_{p/p_0 = 0.1}$$

and are sufficiently accurate for most purposes. One additional feature must be emphasized. The "dead" volume correction is a function of the pressure. For small areas of less than $5 \, m^2 \, g^{-1}$, this correction becomes significant, and precision suffers. However, by using krypton as the adsorbate, better accuracy is possible. Krypton has a p_0 value of 3 torr, compared with 780 torr for nitrogen. The BET range up to $p/p_0 = 0.3$ is covered at pressures 100 times lower, with much less dead volume correction.

In addition to volumetric measurements, gravimetric methods, with microbalances are sometimes used. A typical assembly is shown in Fig. 7.12. Gravimetric methods are preferred with heavier adsorbates such as hydrocarbons.

Also, techniques based on flow systems are used. These have the advantage that high vacuum or pressure measuring devices are not required. They are rapid and accurate enough for most purposes. A typical system is shown in Fig. 7.13. Helium carrier gas containing a known partial pressure of nitrogen passes over a degassed sample at liquid-nitrogen temperature.

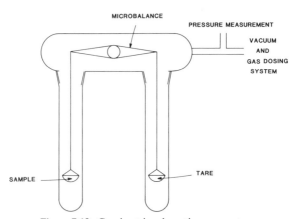

Figure 7.12. Gravimetric adsorption apparatus.

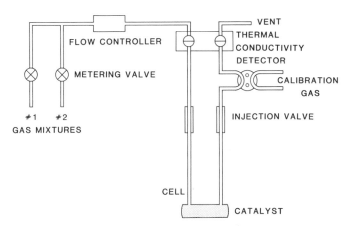

Figure 7.13. Flow system for BET adsorption measurements.

Adsorption is measured from the disappearance of nitrogen, detected with the thermal conductivity cell. No dead volume correction is necessary. This technique is ideal where only single or dual point BET data are desired at low cost.

7.3.5. Pore Size Distribution

Measurement of pore size distribution has now become an essential feature of particle characterization.[216] Questions of pore diffusional resistance, pore mouth poisoning, and deactivation control can only be answered if accurate descriptions of both size and shape are available over the whole pore size range. Historically, macropores have been measured with mercury porosimeters and mesopores with nitrogen adsorption–desorption isotherms.

7.3.5.1. Porosimetry

Some liquids wet solids very poorly. An example is mercury, which exhibits wetting angles, θ, between 112° and 142°. Such a liquid penetrates pores only when forced by a pressure

$$P = -\frac{2S \cos \theta}{r} \qquad (7.10)$$

where S is the surface tension and r the pore radius.[210] For mercury with $S = 474$ dynes cm^{-1} and $\theta = 130°$, the calculated values of P are shown in Table 7.5.

TABLE 7.5. Pressure Necessary to Fill Pores

Pressure (atm)	Pore radius (nm)
1	6×10^3
3.5×10^2	17.5
10^3	6
4×10^3	1.5

Cumulative pore penetration curves are obtained by measuring the volume of mercury forced into the solid at different pressures. Figure 7.14 shows a typical mercury porosimeter and Fig. 7.15 a cumulative penetration curve. Derivative distribution data are also given in Fig. 7.15. Until recently, most porosimeters were limited to pressures of 3.5×10^2 atm. From Table 7.5, the limiting radius is about 17.5 nm. Thus we see why macropores are arbitrarily defined as greater than 15 nm. Higher-pressure equipment is now available, so that, in principle, this technique is feasible throughout the macro and mesopore range.

7.3.5.2. Nitrogen Adsorption

Nitrogen adsorption isotherms, as shown in Fig. 7.9, when extended to p/p_0 values approaching unity, include the region where nitrogen condenses in the pores. Figure 7.16 demonstrates the region for mesopore condensation. This phenomenon is governed by the Kelvin equation, first derived for capillary condensation:

$$r = \frac{2SV \cos \theta}{RT \ln(p_0/p)} \tag{7.11}$$

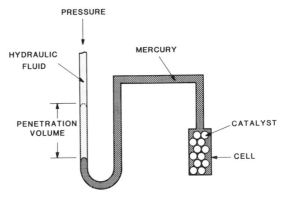

Figure 7.14. Typical mercury porosimeter.

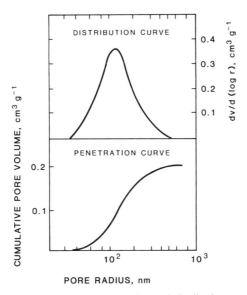

Figure 7.15. Cumulative penetration and distribution curves.

where S is the surface tension of liquid nitrogen, V is the molar volume, θ is the contact angle, R is the gas constant, T is the temperature, and r is the pore radius for condensation at p/p_0. For nitrogen at $-196°C$, r in nanometers is

$$r = -0.956 \ln(p/p_0) \tag{7.12}$$

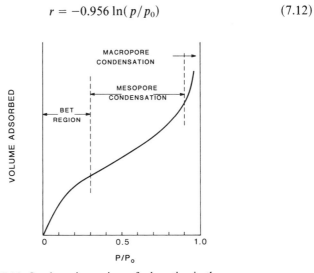

Figure 7.16. Condensation regions of adsorption isotherms.

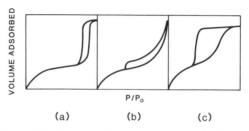

VOLUME ADSORBED

P/P₀

(a)   (b)   (c)

Figure 7.17. Hysteresis shapes in adsorption isotherms.

Equation (7.12) indicates that smaller pores fill at lower pressures. The adsorption curve in Fig. 7.16 is used to construct cumulative pore-filling curves similar to Fig. 7.15 but in the mesopore region. In practice, equation (7.12) must be corrected for the layer of adsorbed nitrogen of thickness t, which effectively decreases the radius,[217]

$$r = t - 0.956 \ln(p/p_0)$$

$$t = -0.605(\ln p/p_0)^{1/3}$$

(7.13)

A complication arises because isotherms display hysteresis, as shown in Fig. 7.17. Much has been written on the origin of these hysteresis curves, which provide information about the shape of pores. There are two extremes. Cylindrical or slit-shaped pores (Fig. 7.18a) give moderate hysteresis, such as Fig. 7.17a. The adsorption branch of the isotherm results from adsorbed layers on the cylindrical walls, which thicken until a miniscus forms. Applicability of the Kelvin equation to such a microscopic system is doubtful. Upon desorption, however, evaporation occurs at the larger miniscus and the Kelvin equation is valid. In this case, pore size distributions from the desorption branch are recommended as more reliable. The other extreme

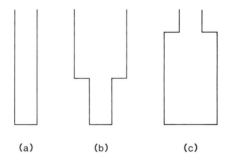

(a)   (b)   (c)

Figure 7.18. Types of pore shapes (a) cylindrical, (b) wide-mouth, (c) ink bottle.

is the "ink-bottle" pore shown in Fig. 7.18c, in which narrow openings access larger volumes. Condensation in the adsorption branch occurs in the large volume. With desorption, lower pressures are necessary to empty the neck, giving the hysteresis curve (c) in Fig. 7.17. This branch is more indicative of aperture sizes.

Usually, intermediate curves, such as (b) in Fig. 7.17, are found. Possibly some type of constricted pore system prevails or perhaps shapes as in (b), Fig. 7.18. This adds some uncertainty to which branch to use. Some discrimination is possible by comparing surface areas from the adsorption and desorption branches, S_{ads} and S_{des}, respectively, and applying the following criteria:

1. If $S_{ads} < S_{BET} \approx S_{des}$, pores are cylindrical and the desorption branch describes the pore size distribution.
2. If $S_{ads} < S_{BET} \ll S_{des}$ severe pore neck constrictions exist and each branch may be used.
3. If $S_{ads} < S_{des} \ll S_{BET}$, considerable micropore volume is present so that neither is accurate.

7.3.6. Diffusivity

Measurement of diffusivity is not normally practiced in characterizing particle properties. This is a pity, since valuable information is available with simple techniques. It is true that the value of measured diffusivities in calculations of effectiveness factors is doubtful. Exact conditions of pressure, temperature, and concentration profiles in the pellet are impossible to duplicate. Diffusion coefficients estimated from rate–particle size curves generally do not agree with direct measurements. However, by comparing calculated to measured diffusivities in simple mixtures, it is possible to estimate the tortuosity factor

$$\tau = D_{calc}\theta / D_{meas} \qquad (7.14)$$

which is not only transferable to more severe conditions but is also an index of texture.

The classical method for measuring diffusivity is the Wicke–Kallenbach technique shown in Fig. 7.19.[21] The catalyst in the form of a cylinder is constrained in a cell which may be as simple as a piece of tygon tubing. More elaborate temperature–pressure cells have been devised. Hydrogen on one side of the cylinder diffuses into nitrogen on the other, and vice versa. This flux is measured and gives D_{eff}. A steady-state technique, this method does not detect diffusion into closed-end pores. Although simple to use, the biggest objection is the need for cylindrical samples. Pellets are

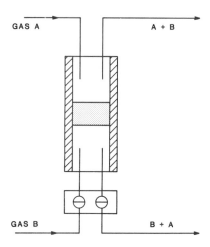

Figure 7.19. Wicke-Kallenback method for measuring diffusivity.

applicable, but extrudates, spheres, irregular particles, and used or distorted particles are not.

A technique that overcomes these limitations is based on the principles of chromatography, as shown in Fig. 7.20.[218] The catalyst, in any form, is packed in a tube. A pulse of argon, injected into helium carrier gas, is

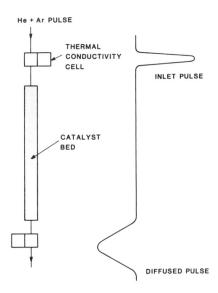

Figure 7.20. Chromatographic method for measuring diffusivities.

TABLE 7.6. Results of Diffusivity Measurement on CoMo/Al$_2$O$_3$ HDS Catalysts

wt % C	S (m^2 g^{-1})	θ	D_{eff} (cm^2 s^{-1})	T dependence
0	270	0.617	0.055	$T^{3/2}$
8.0	219	0.545	0.049	—
9.0	180	0.542	0.036	—
15.3	160	0.297	0.035	—
19.9	49	0.203	0.028	$T^{1/2}$

broadened by diffusion as it passes through the particles. Dimensions of the eluted pulse are used to calculate the van Deempter curve. Measurements at different velocities then lead to an estimation of D_{eff}. Table 7.6 gives results of measurements on carbon-containing hydrodesulfurization catalysts.[218]

The relative change in tortuosity between the fresh and heavily fouled catalyst is only 0.9, whereas the surface area drops by a factor of 5, diffusivity by 2. The most revealing result is the temperature dependence, $T^{3/2}$ for fresh, $T^{1/2}$ for fouled, indicating a change from bulk to Knudsen diffusion. A fouling model was suggested from these results, that is, uniform deposition on the surface, reducing pore diameters evenly, with no preferential blocking of the pores.

7.4. SURFACE PROPERTIES

A catalyst is a surface-active agent. Measurement of surface phenomena in catalysis has occupied scientists since the beginning of research in the field. Surface chemists have borrowed the techniques of colloid chemistry to probe the surface with molecules characterized by various absorption spectroscopies, such as ultraviolet and infrared. Electron microscopy developed into a sensitive tool leading to breakthroughs in understanding surface morphology. But during the last two decades, a revolution in surface technology has occurred. Surface physicists have evolved new generations of high-technology methods to study composition, structure, and interactions on the surface itself. For the first time in catalysis, we now have the opportunity to observe surface phenomena.[219,220]

These techniques often involve photon and electron bombardment and emission, with a vast number of possibilities. A multiple of methods and acronyms have appeared, bringing confusion for the uninitiated. Clean well-characterized surfaces are studied under conditions of ultrahigh vacuum to preserve purity. Research of this type is now beginning to make significant contributions to our understanding of practical catalysts, such as ammonia synthesis. Instrumentation has developed to a point where these methods are used on powders and have become available as routine tools to most organizations. They now assume a position in the array of characterization techniques discussed in this chapter.

We will consider only those methods from surface chemistry and physics that are currently applicable to practical powder catalysts and that are now, or soon will be, readily available to most workers. These cover the characterization of (1) morphology and composition, (2) structure, (3) dispersion, (4) acidity, and (5) activity. The first is included since the size and shape of component crystallites is intimately connected to surface properties. The last factor, activity, must be considered a vital characterization tool during catalyst development.

7.4.1. Morphology

Shape and size are two important features of morphology. Of secondary interest is crystallite size distribution. Information of this nature is necessary for complete knowledge of all catalytic components. It is especially significant for highly dispersed systems. Instrumentation for this purpose includes various types of electron microscopy and x-ray devices.

7.4.1.1. Scanning Electron Microscopy (SEM)

Scanning electron microscopy (SEM) scans over a sample surface with a probe of electrons (5–50 kV). Electrons (and photons), backscattered or emitted, produce an image on a cathode-ray tube, scanned synchronously with the beam. Magnification of 20–50,000 are possible with a resolution of about 5 nm. There is a very high depth of field and highly irregular structures are revealed with a three-dimensional effect.

The SEM is a powerful tool for study of overall topography.[221] Sample preparation is not demanding, so that practical catalysts are easily handled. Resolution limitations, however, restrict the technique to crystallites larger than 5 nm in size. Above this level, crystallite shape, size, and size distributions are easily obtained. SEM investigations have been made on many systems and are useful, not only in crystallite characterization but also in pore structure studies.[203]

7.4.1.2. Transmission Electron Microscopy (TEM)

In transmission electron microscopy (TEM), 100 kV or higher electrons are transmitted through a thin spectrum and the scattered electrons magnified with electromagnetic optics. Images are projected onto fluorescent screens or video detectors with magnifications up to 1,000,000, at better than 0.5 nm resolution under ideal conditions. Modern detectors are amenable to a large amount of computer image interpretation and enhancement. Electron microscopies use sophisticated techniques, such as bright-field, dark-field, or lattice imaging, to enhance information. A new mode of operation, scanning transmission electron microscopy (STEM), provides three-dimensional images for analysis at high resolu ɔn, thus extending the range of SEM into the size region found in highly dispersed catalysts.[203,204]

Disadvantages of this method are (1) high vacuum, (2) sample preparation, and (3) electron beam damage. Images that are recorded and analyzed may not be representative of the average state of the catalyst. Nevertheless, considerable information is available through TEM. One modification, controlled atmosphere electron microscopy (CAEM), utilizes special cells operating at pressures up to one atmosphere and to 100°C. Resolution is reduced to 2.5 nm, but catalytic surfaces operating under dynamic conditions have been studied.[212]

7.4.1.3. X-Ray Diffraction (XRD)

X-ray diffraction (XRD) is mostly used for bulk structure analysis, yet it has one feature which makes it suitable for size determination, if the concentration of the active component is large enough. Figure 7.21 shows how x-ray diffraction lines broaden as the crystallite size decreases. Quantitatively, a number average size is obtained from

$$d = \frac{0.94\lambda}{(B^2 - b^2)^{1/2} \cos \theta} \tag{7.15}$$

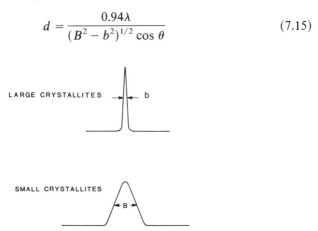

Figure 7.21. X-ray diffraction line broadening of dispersed crystallites.

where B is the peak width for a diffraction line at angle θ, b is the value for a well crystallized specimen, and λ is the wavelength.[223]
Reliable data are only possible down to 5 nm since smaller crystallites give such broad lines that sensitivity decreases. An advantage to this method is that the sample may be heated and exposed to reactive atmospheres, and changes in size during dynamic conditions are observable. Appropriate line profile analysis measures crystallite size distributions.[224] When perfected, this approach will be a valuable adjunct to morphology characterization.

Small angle x-ray scattering (SAXS) has been used in the past to study pore size distributions in amorphous materials. The method gives good results but is not now widely practiced.[223]

7.4.1.4. Extended X-Ray Absorption Fine Structure (EXAFS)

This is a relatively new tool that shows great promise. X-rays, when absorbed, transmit photon energy to inner electrons, which then escape from the atom. Interaction between these electrons and neighboring atoms produce fine structure in the x-ray absorption edge, giving information on coordination numbers and interatomic distance. Unfortunately, high intensity x-rays from synchrotons are necessary, so that the technique is not readily available. Nevertheless, valuable information on surface environments, not available from other sources, is beginning to appear.[1]

7.4.1.5. Auger Electron Spectroscopy (AES)

This is a tool that is truly surface sensitive. The basic AES process is illustrated in Fig. 7.22. Electrons (1–5 kV) generate holes in core electron levels, although soft x rays give identical effects. A valence (or core) electron

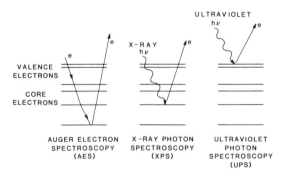

Figure 7.22. Electron spectroscopies for surface analysis, Auger electron spectroscopy, X-ray photon spectroscopy, and ultraviolet photon spectroscopy.

decays into the hole releasing enough energy to eject another electron. This is the Auger electron, with a kinetic energy characteristic of its atomic or molecular energy level, i.e., representative of the atom. All elements except H and He are detectable to a depth depending on the surface and incident energy. Probing from 0.5–1.0 nm is possible. Certainly, measurements of up to ten layers can be controlled. Quantitative analysis and chemical (environmental) information is not as reliable as with XPS (see below) and the electron beam can induce damage. Nevertheless, it is a rapid method for probing surfaces, which, when combined with Ar^+ bombardment to "scrape" surface layers, results in information on depth profiles.[225]

7.4.1.6. X-Ray Photon Spectroscopy (XPS)

In x-ray photon spectroscopy (XPS), a soft x-ray photon (170 eV) ejects any electron with less binding energy. When properly detected and counted, the electrons give a spectrum characteristic of the element. Excellent sensitivity is possible with probing depths of 1–20 layers. X-ray photon spectroscopy spectra are sensitive to the chemical environment, so that analysis of the "chemical shift" leads to information on structure and differences in surroundings are detected. The technique has poor lateral resolution, with sample areas from 0.5 to 100 mm^2, and profiling with Ar^+ is not as easy as with AES.[226]

7.4.1.7. Ultraviolet Photon Spectroscopy (UPS)

Ultraviolet photon spectroscopy is identical to XPS, except that lower-energy photons (30 eV) are used, limiting analysis to valence electrons. This is useful in the analysis of adsorbed states but is no advantage over XPS for surface characterization.

7.4.2. Structure

Identification of surface phase structure is best accomplished with either SEM or TEM, which provide facilities for electron diffraction of selected surface regions. An individual crystallite or surface region is selected, and its structure analyzed in the same way as bulk structure is determined with XRD.

This is complementary to XPS and UPS spectra that reveal compound types through chemical shifts, usually with "fingerprinting." Auger electron spectroscopy (AES) is also used but interpretation is complicated.

An example is the study of sulfided $CoMo/Al_2O_3$ hydrodesulfurization catalysts, showing the presence of discrete crystallites of Co_9S_8 and MoS_2, dispersed over a sulfided molybdenum layer and superimposed on a substrate of $CoAl_2O_4$. Further identification of Co ions in the molybdenum layer and MoS_2 microcrystals was possible.[227]

7.4.3. Dispersion

Dispersion is defined as the fraction of active atoms in surface positions. Higher dispersions are achieved either as monolayers of atomically dispersed material or as very small crystallites. Both exist in practice. In either case, measurement of surface concentration is most easily performed by quantitative probing with selective molecules. There are three common approaches: measurement of (1) chemisorption isotherms, (2) reaction titration, and (3) poison titration.

Before discussing these techniques, it is perhaps wise to comment on the practice of calculating active surface areas from crystallite size measurements. This is unreliable for several reasons. First, instrumental techniques such as SEM, TEM, and x-ray line broadening all have their limitations. Monolayer-type dispersions are not detected and ultrasmall crystallites are below the range for accurate determination. Second, shape must be assumed in the calculations. Theoretical and experimental evidence indicates that the most stable shapes are spherelike cubo-octahedra. However, depending on the interaction with the support, crystallites may exist as full spheres, hemispheres, or intermediate structures. For very small sizes, this cannot be resolved easily with TEM. Also, a distribution of sizes and shapes may exist. Calculation of the surface from size measurements is, at its best, an indication of an upper limit.

Third, there is increasing evidence that not all of the active surface is accessible.[228] Figure 7.23 illustrates possible reasons for this. Small or growing crystallites may be completely or partly trapped in pores of comparable size. Crystallites may be encapsulated or partially covered by patches of support or compounds of the support and the active component. In either case, calculation of surface from size yields too high a value.

It is better not to calculate dispersions but to measure them. A bonus results if both surface area and size distribution are available and the degree of accessibility can be determined.[228]

7.4.3.1. Chemisorption Isotherms

The essential features of chemisorption are given in Table 7.3. For dispersion measurements, the most important are specificity and monolayer coverage. The approach is to measure an adsorption isotherm under conditions that give predominantly chemisorption with full monolayer coverage,

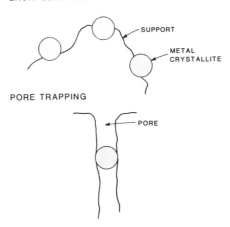

Figure 7.23. Inaccessibility of metal crystallites on supports due to encapsulation and pore trapping.

using an adsorbate that is specific to the active component and not to any other. Metals are especially susceptible to this method because of distinctive chemisorption differences with support oxides. Most applications in recent years have been supported metals. Table 7.7 gives a partial list of selective adsorbates forming monolayers on different metals.

Once a suitable gas has been identified, two criteria must be satisfied.[223] First the monolayer must form rapidly. Well-reduced dispersions of noble metals on alumina satisfy this requirement, as indicated in Fig. 7.24. Monolayer volumes are easily detected. Other cases are not so evident. In Fig. 7.25, for example, hydrogen adsorption on Ni/SiO_2 does not saturate.[228] Monolayer coverage is assumed to take place very rapidly with reversible adsorption occurring on sites deeper in the surface or at the

TABLE 7.7. Metals and Gases That Form Monolayers

| Metals | Gases[a] | | | | | |
	N_2	H_2	O_2	CO	C_2H_4	C_2H_2
W, Mo, Zr, Fe	+	+	+	+	+	+
Ni, Pt, Rh, Pd	−	+	+	+	+	+
Cu, Al	−	−		+	+	+
Zn, Cd, Sn	−	−	+	−	−	−
Pb, Ag, Au	−	−	−		+	+

[a] +, gas forms monolyaer; −, gas does not form monolayer.

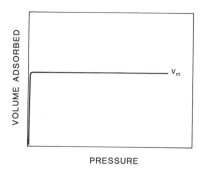

Figure 7.24. Hydrogen chemisorption on Pt/Al$_2$O.

interface with the support. Extrapolation back to zero pressure is necessary to determine the monolayer.

Second, the stoichiometry of adsorption must be known in order to calculate surface concentrations. This is extremely difficult to establish. Hydrogen, for example, adsorbs as M–H species over crystalline planes. This is confirmed with parallel hydrogen chemisorption and BET measurements on nonsupported nickel. However, the stoichiometry of the bond M–H$_n$ appears to increase for low coordination sites (see Chapter 3) so that overall values for very small crystallites may be greater than one. Carbon monoxide is even more troublesome. Several modes coexist; linear Ni–CO, bridged, Ni$_2$CO, and subcarbonyl, Ni(CO)$_2$, so that some assumptions are inherent in its use.[223]

In spite of these difficulties, H$_2$ and CO chemisorption are favorites for dispersion measurements. Table 7.8 gives typical results[229] and Table 7.9 lists other successful applications.

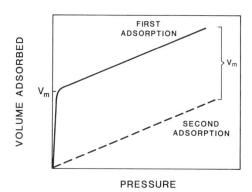

Figure 7.25. Hydrogen chemisorption on Ni/SiO$_3$.

TABLE 7.8. Typical Results for SiO_2-Supported Noble Metals

Catalyst	H/M	CO/M	Surface area $m^2 g(metal^{-1})$
10% Pd	—	0.11	47
10% Rh	0.48	0.44	214
5% Rh	0.53	0.52	237
5% Ru	0.27	0.31	121
10% Ir	0.78	0.58	186

TABLE 7.9. Applications of Surface Area Measurement by Chemisorption

Metal	Adsorbate	Reference
Fe	CO	234
Ni, Co	H_2, CO	231
Cu	CO	235
Ag, Au	O_2	232
Pd	CO	236
Pt, Ru, Rh, Ir	H_2, CO	229
Ru	O_2	233

Chemisorption characterization of nonmetals has not been so fully exploited. A few examples from the literature are given in Table 7.10. There is no reason why the method should not be equally applicable to oxides and sulfides, providing the necessary criteria of specificity, monolayer coverage, and known stoichiometry are met.

The techniques for chemisorption measurement are the same as those discussed for total adsorption and include volumetric, gravimetric and flow methods. Each case is different, owing to the relative amounts of physical adsorption that must be taken into account.[242]

TABLE 7.10. Applications of Chemisorption to Nonmetals

Catalyst	Adsorbate	Reference
MoO_3/SiO_2	O_2	237
Cr_2O_3/Al_2O_3	NO	238
NiO/Al_2O_3	NO	239
Fe_2O_3/Al_2O_3	NO	240, 241
MnO/Al_2O_3	O_2	242
$K_2O/Fe-Al_2O_2$	CO_2	234

7.4.3.2. Reaction Titration

In cases not readily amenable to direct chemisorption, titration of specific sites is possible with probe reactions. The reaction must be an irreversible interaction between a gas and surface site, such that only one event occurs per site. The first example of this technique was the titration of Pt-O sites.[243] The surface of supported platinum is first oxidized to form no more than a monolayer. Then follows reaction with hydrogen

$$Pt-O + \tfrac{3}{2}H_2 \rightarrow Pt-H + H_2O \tag{7.16}$$

Uptake of hydrogen is increased by a factor of 3 over hydrogen chemisorption. This results in greater precision for catalysts with low loading.

Another example is copper surface area measurement in Raney Cu and Ni-Cu alloys. When reacted with N_2O, most conveniently in a pulse apparatus, Cu reacts as follows:

$$N_2O(g) + 2Cu_s \rightarrow N_2(g) + (Cu_s-O-Cu_s) \tag{7.17}$$

Measurement of eluted nitrogen pulses gives the surface copper concentration directly.[244] The same technique has been used with supported silver catalysts.[245]

7.4.3.3. Poison Titration

Poison titration is a convenient way to measure the concentration of active sites. The best procedure is to use a simple pulse reactor, such as that in Fig. 7.26. Pulses of a poisoning agent are injected between reactant pulses. If all the poison adsorbs irreversibly, then activity declines with each pulse. Typical results are shown in Fig. 7.27, in which hydrogen sulfide poisons metal sites.[246] Extrapolation of the activity curve to zero gives the amount of poison necessary to neutralize the active sites. A knowledge of surface stoichiometry is necessary to proceed further. For example, in Fig. 7.27 the assumed ratio was two nickel for each sulfur.[246] This technique has the potential for innovative application to many systems.

7.4.4. Acidity

The nature of acidity on insulating oxides is discussed in Chapter 3. Important factors are (1) type of acidity, (2) acid strength, and (3) distribution of acid strengths. It is not easy to distinguish between Lewis or Bronsted

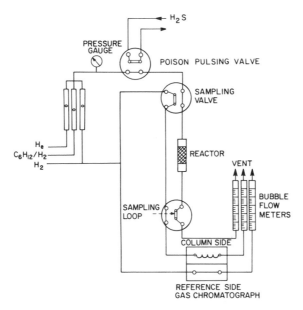

GAS HANDLING SYSTEM

Figure 7.26. Pulse reactor apparatus for poison titration measurements using hydrogen sulfide.

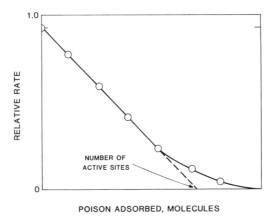

Figure 7.27. Typical results for the titration of metal sites by a poison.

sites. The most successful approach is to adsorb a strong base, such as pyridine, and measure the intensity and wavelength of resulting infrared adsorption bonds.[247] Table 7.11, for example, shows the results of a study on Faujasite Y catalysts ion exchanged with different cations.[248] Lewis acidity increases dramatically as the polarizing power of the exchanged cation increases. Bronsted acidity shows a moderate increase, except with hydrogen exchange, which is almost completely Bronsted.

Similar studies have shown that Bronsted acidity is dominant in most acidic solids. Differences in acid strengths are observable, but quantitative determinations are difficult and have not resulted in accepted methods for acid distribution measurement. Methods that have gained wide acceptance are (1) nonaqueous titration and (2) base chemisorption.

7.4.4.1. Nonaqueous Titration

The most direct method is to adsorb an indicator on the catalyst in suspension with a nonpolar solvent. The indictor has a known pK_a for the acid–base color change. A base is then added until the acid–base end point is observed. The amount of base indicates the number of acid sites with strengths less than the pK_a of the indicator. By using a range of indicators with different pK_a's, the distribution of acid strengths is determined. End points are detected either by visible color changes or with spectrophotometry. For visible detection, the Hammett indicators used are given in Table 7.12.

Typical results are shown in Fig. 7.28 for fresh and steamed silica-alumina cracking catalyst.

Some workers report that Hammett indicators give unreliable results at higher values of pK_a. Better data are obtained with fluorescent indicators

TABLE 7.11. Acidity Type by I.R. Pyridine Measurement

Faujasite	$1545\ cm^{-1}$ (B)[a]	$1490\ cm^{-1}$ (B + L)[b]	$1438\ cm^{-1}$ (L)[c]
Na^+	0	0.27	1.2
Li	0	0.29	2.1
Ca^{2+}	0.33	1.0	4.1
Mg^{2+}	0.41	1.3	3.0
Cd^{2+}	0.57	1.8	6.8
H^+	7.0	—	0.8

[a] $1540\ cm^{-1}$, pyridinium ions (Bronsted, B).
[b] $1490\ cm^{-1}$, pyridinium + coordinately bound pyridine.
[c] $1438\ cm^{-1}$, coordinately bound pyridine (Lewis, L).

TABLE 7.12. Hammett Indicators Used in Acidity Measurement

Indicator	Acid color	Base color	pK_a	Equivalent % H_2SO_4
Brom thymol blue	Yellow	Blue	−6.8	
p-Ethoxychrysoidin	Red	Yellow	+5.0	
4-Phenylazo-1- naphthalamine	Red	Yellow	+4.0	
Aminoazoxylene	Red	Yellow	+3.5	
Benzene-azo- dimethylaniline	Red	Orange	+3.2	
2-Amino-5-azotoluene	Red	Yellow	+2.0	
Benzeneazo- diphenylamine	Purple	Yellow	+1.5	0.02
p-Nitrobenzeneazo diphenylamine	Violet	Orange	0.43	
p-Nitrodiphenylamine	Red-Orange	Orange	−2.1	40
Dicinnamalacetone	Red	Yellow	−3.0	48
Benzalacetophenoue	Yellow	Colorless	−5.6	71
Anthraquinone	Yellow	Colorless	−8.2	91

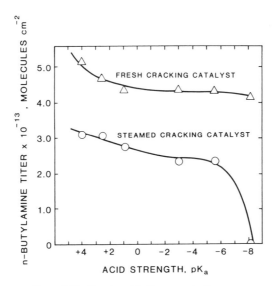

Figure 7.28. Results with Hammett indicators.

TABLE 7.13. Improved Fluorescent Indicators[a]

Indicator	Wavelength (m)		pK_a
	Acid	Neutral	
5,7-Dimethyl-1-2-benzacridine	480	400	+5.5
6,9-Dichloro-2-methoxyacridine	520	440	+3.0
Dimethyl POPOP	470	420	+0.5
POPOP	460	410	−1.8

[a] Reference 249.

in which the end point is found spectrophotometrically.[249] Table 7.13 lists some of these indicators and Fig. 7.29 shows the improved results.

Although these titration methods are in wide use throughout the industry, there is considerable doubt concerning their validity. Neutralization on the surface is not in equilibrium, and different results are obtained with alternate bases, concentrations, and time of experiment. At the best, it is a relative procedure in which all conditions must be constant at specific values. Research is continuing into these phenomena, and it is anticipated

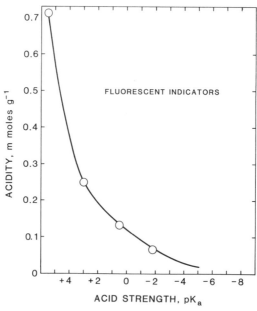

Figure 7.29. Results with fluorescence indicators on cracking catalysts.[249]

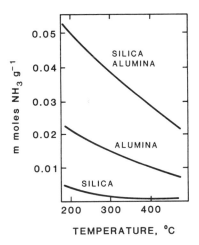

Figure 7.30. Ammonia desorption to measure acidity.

that new techniques, possibly involving measurement of adsorption heats of bases with different strengths, will be forthcoming.

7.4.4.2. Base Chemisorption

Strong bases such as NH_3 and pyridine ($pK_a = +4.8$ and $+5.2$) adsorb on acid sites with a strength of adsorption proportional to the acid strength. Qualitatively this is shown in Fig. 7.30. The ordinate is a measure of the number of sites, the abscissa the acid strength, since higher temperatures are necessary to desorb more strongly held molecules. The isobars are then acid strength distribution curves and give at least relative differences between samples.[250] Measurements are made at one temperature or over a range of temperatures, using gravimetric or pulse methods. A rapid analysis is obtained by using a temperature programmed desorption (TPD) modification of a TGA. This approach, although relative, allows comparisons to be made for a given series of catalysts.

7.5. ACTIVITY

Throughout catalyst development, "activity" is a key parameter in design, selection, and optimization, and yet the term means different things at different times to different people. In this section, we consider commonly used expressions for activity, what they mean, when to use them, and how to measure them.

Activities fall naturally into two types: (1) kinetic and (2) practical. Kinetic expressions of various kinds are essential in research and process reactor design. Rate equations and associated parameters direct the chemist toward a better understanding of mechanisms, from which improved catalysts are designed, and the engineer toward models, from which processes are optimized. Considerable care and accuracy must go into measurement of these quantities. Practical activities have limited purposes, for example screening of catalyst candidates, optimization of preparational parameters, determination of process parameters, deactivation studies, diagnosis of malfunctioning catalysts, and product quality checks on new catalyst charges. Usually, relative results are sufficient, with speed and productivity the dominant factors.

7.5.1. Kinetic Activities

Kinetic activities are based on the prevailing rate equation

$$-\frac{1}{V}\frac{dC_A}{dt} = k_v f(C_A) \tag{7.18}$$

with

$$k_v = k_{v,o}\, e^{-E/RT}$$

where

$$-\frac{1}{V}\frac{dC_A}{dt}$$

is the rate per unit volume of either reactor, V_r, fluid, V_f, or catalyst V_c; k_v is a rate constant on a volume basis; $f(C_A)$ is a rate equation concentration term, which may be simple order or complex with adsorption terms; $k_{v,0}$ is a preexponential term; and E is the activation energy. Equation (7.18) suggests three ways to express activity: as a rate, a rate constant, or an Arrhenius constant.

7.5.1.1. Reaction Rates

This is the most direct and specific. In catalysis, it is usual to express rate in terms of catalyst volume, V_c, but alternate forms are

$$-\frac{1}{W}\frac{dC_A}{dt}\text{—moles per unit time per unit mass of catalyst} \tag{7.19}$$
$$\text{(specific rate)}$$

and

$$-\frac{1}{S}\frac{dC_A}{dt}\text{—moles per unit time per unit surface, total} \quad (7.20)$$
$$\text{or active (areal rate)}$$

An adaptation of rate per unit surface now in common use is the turnover number, N_t, defined as the number of molecules reacted per site per second. Although appealing in its molecular simplicity, the turnover number should be used with caution, since it requires a knowledge of the surface area under reaction conditions and the stoichiometry or structure of the active site. Surface area is difficult to measure. The most common approach is to find the surface area of the fresh catalyst in a separate experiment, where activation conditions may not be exactly reproduced. Next, the structure of the active site is needed to relate surface area to site density. This is the most elusive property in catalysis and is the subject of much research. It is not an overstatement to say that there are very few reactions where we can even approximate these structures. Perhaps in the future, innovative methods will open a way to use these concepts. In the meanwhile, it is better to represent rates on the basis of a measurable and known property, such as volume, mass, or surface area.

Measurements of rates at different concentrations and temperatures lead to empirical rate equations. If, however, a series of catalysts is to be compared, each must be expressed at a specified temperature and concentration. Initial rates at zero process time are the most reliable. In the case of deactivation or surface conditioning by the reactant, rates should be measured at some standard process time. This must be established carefully and rates interpolated, but never extrapolated, from rates measured over the complete range.

7.5.1.2. Rate Constants

Using a rate constant to express activity avoids the problem of specifying concentrations, but not temperatures. The exact form of the rate equations, however, must be known. This involves extensive experiments, at least initially. If the function of the rate equation is available with sufficient confidence, then subsequent measurement of rate constants requires only a few data points. Rate constants may be expressed in a variety of units, with respect to volume, mass, or surface area. An example for a first-order reaction is given in Table 7.14.

A warning on the use of rate constants should be stated at this point. A common practice in catalysts is to fit kinetic data to the power rate expression

TABLE 7.14. Rate Constants for First-Order Kinetics

Rate constant	Unit parameter	Form	Units
k_s	surface	k_s	$\mathrm{cm} \cdot \mathrm{s}^{-1}$
k_w	mass of catalyst	$k_s S_g$	$\mathrm{cm}^3 \, \mathrm{g}^{-1} \, \mathrm{s}^{-1}$
k_{v_c}	volume of catalyst	$k_s S_g d_p$	s^{-1}
k_{v_R}	volume of reactor	$k_s S_g d_p (1 - \varepsilon)$	s^{-1}
k_{v_f}	volume of fluid	$k_s S_g d_p (1 - \varepsilon)/\varepsilon$	s^{-1}

$$\text{rate} = k_{\text{obs}} P_A^n P_B^m \tag{7.21}$$

The activities of series of catalysts are represented with k_{obs} values. Values outside the ranges of concentrations and temperatures are often found by extrapolation. This is a very dubious procedure. Equation (7.21) is in reality an approximation of a Langmuir–Hinshelwood expression of the type

$$\text{rate} = \frac{k_s K_A K_B P_A P_B}{(1 + K_A P_A + K_B P_B)^2} \tag{7.22}$$

The exponents n and m depend on the values of $K_A P_A$ and $K_B P_B$, which in turn are temperature dependent. The rate constant so derived is

$$k_{\text{obs}} = k_s K_A^n K_B^m \tag{7.23}$$

where n and m are concentration and temperature dependent. Expressions such as equation (7.21) should only be used over the limited range in which measurements were made. Comparisons from one catalyst to another are very uncertain.

7.5.1.3. Arrhenius Parameters

On occasion, the parameters, $K_{v,0}$ and E, are used to specify activity. This method is only used when it is certain that one or the other is constant, in which case, why not use the complete rate constant? A constant value of $K_{v,0}$ or E cannot be assumed since there are many cases in catalysis of compensation effects, where $K_{v,0}$ increases as E decreases or vice versa.[251]

7.5.2. Experimental Reactors

It is now appropriate to introduce the topic of experimental reactors used to measure kinetic rate data. Much has been written on this subject and many reviews are available.[86,252,253,254] Here we shall consider only those distinguishing features important for proper use. The correct reactor should be selected for the task at hand. It is not sufficient to use an apparatus merely because it is available. Will it yield necessary data without complicating the interpretation? Carefully assess the requirements of the experiment and keep the reactor as simple as possible. Avoid the temptation to build versatility, for experience has shown that the more functions a piece of equipment performs, the greater the probability of malfunction. Limitations must be recognized, however, and expectations clearly defined.

Most experimental reactors have one common feature—small size. Catalyst beds are small, usually less than 1 g, perhaps mixed with inert material to give reasonable volumes (1-2 cm^3). This is a distinct advantage. Catalyst costs are low because only small amounts need be prepared. Heat transfer and control difficulties are minimal and temperature is accurately measured. Auxiliary equipment such as tubing, valves, pumps, and gauges, are low cost and easily fabricated. Reagent supplies are reasonable and available from easy storage. Space is conserved. Microanalytical equipment exists for most applications. There are, however, compensating disadvantages. With small catalyst beds, there are reproducibility problems, especially with laboratory preparations. Small beds are much more sensitive to low concentrations of poisons, so that reactants must be thoroughly cleaned to ultrapure levels. The catalyst is often used as a powder, since reactor diameters are low. Although pore diffusion problems are minimized, pressure drop difficulties arise. With low flow rates, the particle Reynolds number, Re_p, is small and the flow regime ill defined. Values of Re_p greater than 25-50 are necessary for turbulent flow; below that it is laminar. Many laboratory reactors have values less than 1-10, where external diffusion correlations are uncertain. However, if these objections are overcome, laboratory reactors offer the most rapid, least expensive, and simplest way to accumulate rate data.

There are four major types (1) tubular, (2) gradientless, (3) batch, and (4) pulse.

7.5.2.1. Tubular Reactors

A typical tubular reactor system is shown in Fig. 7.31. The reactor itself is made of glass, quartz, or stainless steel, depending on the temperature and pressure. The diameter and length should not be less than that required to give tube to particle diameter ratios of 5-10 and length to particle diameter ratios of 50-100. Auxiliary flow and measuring equipment is shown in Fig. 7.32.

The reactor is operated either in an integral or differential mode. As an integral reactor, initial concentrations of reactants are adjusted and the

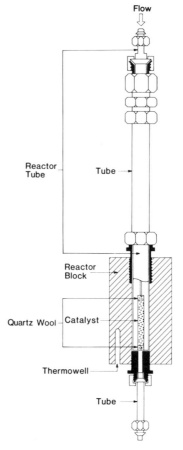

Figure 7.31. A typical metal tubular reactor.

space velocity, F/W, regulated to give a series of conversions. The analysis follows by fitting rate equations to the design equation

$$\frac{W}{F} = C_{A0} \int_0^{X_A} \frac{dX_A}{kf(C_A)} \qquad (7.24)$$

to verify $f(C_A)$ and find k. The initial concentration of A is C_{A0} and X_A is the fractional conversion. This approach is best for simple order rate equations. There are too many constants with Langmuir-Hinshelwood expressions.

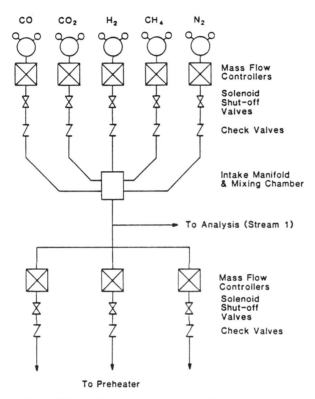

CO   CO₂   H₂   CH₄   N₂

Mass Flow
Controllers

Solenoid
Shut–off
Valves

Check Valves

Intake Manifold
& Mixing Chamber

To Analysis (Stream 1)

Mass Flow
Controllers

Solenoid
Shut–off
Valves

Check Valves

To Preheater

Figure 7.32. Equipment for the control of multiple reactors.

In the differential mode, conversion is maintained at low values of less than 5%, while initial concentrations are varied over as wide a range as possible. Under these conditions

$$\text{rate} = X_A \left(\frac{F}{W} \right) C_{A0} \tag{7.25}$$

and postulated rate equations are checked directly. Complex rate equations adapt well to this procedure, but low conversions require good analytical precision.

Precautions must be taken to ensure that external or pore diffusion resistance is not significant. This may be done either by calculation, as discussed in Chapter 1, or with experimentation in which conversion is checked against linear velocity and particle size. It is also wise in making extended experimental runs to return to some standard experimental condition at the end of the run. If results are the same, this ensures that no deactivation has taken place. If activity has declined, then measured rates may be "corrected" back to the fresh state.

7.5.2.2. Gradientless Reactors

These are reactors that operate in a well-mixed condition. For catalytic systems this is achieved with Carberry or Berty reactors as shown in Fig. 7.33.

Concentrations remain constant throughout the reactor volume, and mixing helps in temperature control. Conversions may be quite large, so that rate precision is good, but interpretation is a simple application of equation (7.25). Another advantage is that pelleted or extruded particles can be used directly. Figure 7.34 demonstrates results with commercial-sized particles.

Checks are necessary to ensure no external diffusion problems occur and that well-mixed conditions prevail. This can be done by changing the speed of the mixer.

Similar effects are achieved with recirculation reactors, in which at least 90% of the product is recirculated to the reactant so as to maintain essentially constant concentrations over the bed. These reactors are,

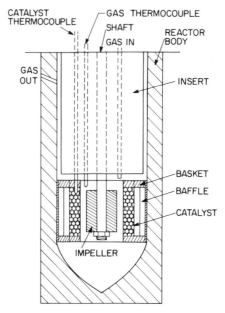

Figure 7.33. A well-mixed basket reactor.

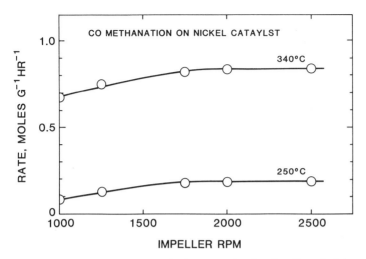

Figure 7.34. Typical results with a well-mixed reactor, showing the effect of mixing speed.

however, difficult to design for many small applications, especially at high temperatures.

7.5.2.3. Batch Reactors

Batch reactors are not common in experimental catalysis since flow systems are so simple. They are most often found with liquid phase reactors using slurried catalysts. Good mixing is essential and checks should be made to eliminate external diffusion problems. Also it is necessary to vary catalyst loading during the experiment as a means of detecting vapor–liquid interfacial effects.[21]

Gradientless and recirculation gas phase reactors may also be operated in a batch mode by eliminating flow, but there is usually little advantage to this in catalyst development.

7.5.2.4. Pulse Reactors

These are rapid, simple reactors that offer economy of feed and provide protection against process deactivation. A typical scheme is given in Fig. 7.26. The operation is transient so that concentrations change over the surface. Kinetic interpretations are difficult for any but first-order rate equations.[255] Pulse reactors do have utility in poison titration measurements, in determining activities of the "fresh" surface, and for following surface conditioning.

7.5.3. Practical Activities

Practical activities are used when catalyst comparisons only are required. Measurements are faster and less demanding. The same types of reactor discussed above are used or larger units may be required. The principal function of pilot unit operations is to explore real feeds, process conditions, and deactivation, but practical activities are usually measured as these change. Three types are encountered; measurement of (1) conversion at fixed space velocity, (2) space velocity for a fixed conversion, and (3) temperature for a fixed conversion.

7.5.3.1. Conversion at Fixed Space Velocity

Catalysts are compared by measuring the conversion at a fixed space time. Consider the conversion space time curves in Fig. 7.35. At the selected space velocity, catalyst A is better than B because the conversion is higher. Feed composition, temperature, pressure, and process time should be constant for the test to be meaningful. The best way to ensure this is to run the samples in parallel reactors with the same feed. Figure 7.36 shows a system designed for this purpose. Three catalysts are measured simultaneously, with microprocessor control to ensure identical and reproducible conditions. One of the samples could be a standard against which the others are checked.

This approach is relative only and has no kinetic meaning. Indeed, the absolute ratio between two activities depends on the space velocity used and has no meaning at other values.

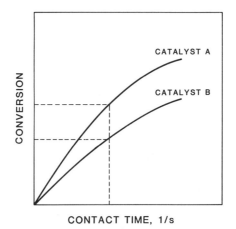

Figure 7.35. Conversion versus space time tests.

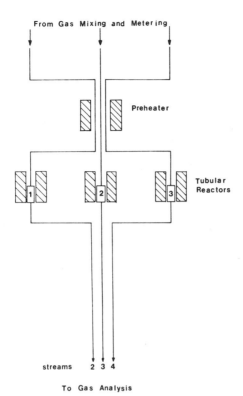

Figure 7.36. Typical equipment for multiple testing of catalysts.

7.5.3.2. Space Velocity for a Fixed Conversion

Figure 7.37 demonstrates the principle in which conversion is maintained constant and space velocities compared. Catalyst A is better than B because lower space times yield the same conversion. This is the best practical method but not the easiest to perform. If the reaction rate equation is simple order and

$$\frac{W}{F} = \frac{1}{S} = \frac{C_{A_0}}{k} \int_0^{X_A} \frac{dX_A}{f(X_A)} \tag{7.26}$$

then

$$\frac{1}{S} \propto \frac{1}{k} \tag{7.27}$$

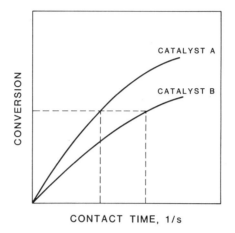

Figure 7.37. Constant conversion tests.

for a constant X_A, so that

$$\frac{S_A}{S_B} = \frac{k_A}{k_B} \tag{7.28}$$

Ratios of rate constants are obtained without kinetic analysis.

7.5.3.3. Temperature for Fixed Conversion

A series of conversion versus temperature curves is shown in Fig. 7.38 to demonstrate this method. Conversion is maintained constant by changing

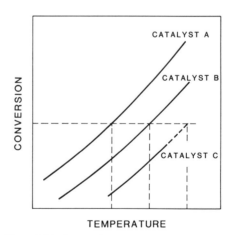

Figure 7.38. Conversion versus temperature tests.

the temperature. Catalyst A is better than B because it gives identical conversion at lower temperatures. This is the most dangerous of all practical methods. Catalyst C may not reach the specified conversion within limits of the apparatus. It is tempting to extrapolate as shown. This could be misleading, since kinetics may change at higher temperatures. Also, if compensation effects exist, curves for A and B cross, and opposite conclusions result depending on the conversion level used. Although useful for limited temperature ranges, this method has no kinetic significance. It is often used to follow deactivation in pilot plants.

7.6. SOME FINAL THOUGHTS

We have seen that a large number of techniques exist to characterize catalyst properties and measure activities. Not all of these will be useful in a given application. The experimenter must be discriminating and select the tests that yield the most information with accuracy, speed, and economy. To do this, it is necessary to know precisely the objectives of the investigation, the capabilities and limitations of each technique, and the expected value of the results.

CATALYST DEACTIVATION
How They Fail

8.1. PRELIMINARY REMARKS

Ideally catalysts should last forever. In reality, they die with use, victims of many diseases that rob them of their ability to function. Some go quickly in a matter of minutes, others last for up to ten years, but ultimately they all succumb. Catalyst and process designers together can do much to prolong lifetime, but as one illness is cured the victim falls prey to another. In this chapter we examine the cause and effect of deactivation, with emphasis on catalyst modifications that prove effective in combating it. Only those features that are important in commercial processing are considered, and numerous review articles should be consulted for additional background.

8.2. EFFECTS OF DEACTIVATION

Deactivation is loss of activity. The rate per unit volume of reactor is given by

$$\frac{1}{V_R}\frac{dC_A}{dt} = -\left(\frac{N_t}{A}\right)C_s S_g d_B \tag{8.1}$$

where N_t is the turnover number, A Avogadro's number, C_s the surface concentration of sites, S_g the surface area per gram, and d_B the packing density. We shall identify later the factors that influence each of these parameters. Usually N_t is invariant and deactivation decreases one or more of the other parameters. This is an oversimplification, since N_t may also be changed by the deactivating process, yet it serves a useful model for discussion. For the present, we shall consider only the effect of loss of total activity as measured by the rate or some derivative such as rate constant, conversion, and so on.

In laboratory measurements, deactivation is troublesome and misleading if not identified. A clean catalyst surface begins to deactivate as soon as it encounters reactant molecules. Usually, this initial phase is so fast that it may not be observed by the time the first measurement is made. Curve (A) in Fig. 8.1 shows how this loss develops. Steady state is achieved after some finite time. The reason for this is usually permanent deactivation of very active sites that play no further role in the catalysis. Such effects are more obvious with pulse reactors. For example, during cyclopropane hydrogenolysis in a pulse reactor, the first pulse of cyclopropane reacted completely by hydrogenolysis.[61] In successive pulses, the extent of hydrogenolysis decreased as carbonaceous residues poisoned the most active sites. Finally, a steady state distribution between hydrogenation and hydrogenolysis was achieved.

In Fig. 8.1, curve (B) is an example in which activity increases to its steady value. There is also some speculation that this is a process in which an "overlayer" of some intermediate compound, perhaps carbon, deposits and activity develops with interaction between normal surface sites and this

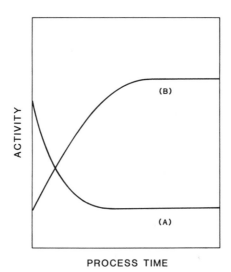

PROCESS TIME

Figure 8.1. Initial conditioning of the catalyst. (A) Destruction of very active sites, (B) development of active sites.

overlayer.[256] Much work remains to be done before these concepts are confirmed.

Nevertheless, trends such as those shown in Fig. 8.1 are observed and must be regarded as part of a deactivation process in which the surface is "conditioned." This should not be confused with activation, which is a well-defined phase change, as in reduction or sulfiding.

Laboratory rate measurement should start with the conditioned catalyst at steady state. Rapid deactivation thereafter confuses these measurements, especially with the type of time-consuming experiments used to determine rate equations. In such cases, it is wise to return to initial conditions at the end of each run and check for differences. Whenever possible, deactivation should be avoided or appropriate corrections made in some reasonable manner. This becomes more difficult when deactivation is rapid.

Conditioning is also observed in process units, but is either overlooked or considered part of the start-up. Process deactivation follows a pattern shown in Fig. 8.2. Activity declines with process time at a rate that depends on prevailing conditions. It may be gradual, follow a well-defined schedule, or be very rapid. Examples are naphtha hydrotreating with lifetimes of up to several years, and catalytic cracking, where substantial deactivation occurs in only minutes.[257] Ultimately, deactivation reaches a point where conversion or other conditions are below design specifications and the

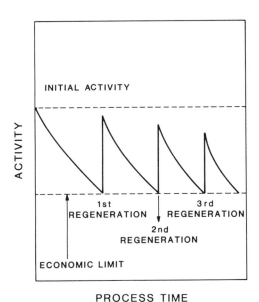

Figure 8.2. Activity decline in process units.

catalyst must be either replaced or regenerated. Regeneration is a treatment in which activity is returned. In practice, initial activity is not always restored, due to a small permanent secondary deactivation. Further processing continues for another cycle, until regeneration is necessary again. Finally, regeneration becomes unproductive, and catalyst replacement is indicated.

A sudden change in deactivation rate is seen when catastrophic upsets occur. This may be caused by unit malfunction, error, or unexpected changes in feed properties. Many models have been used to predict deactivation rates, but the most general and useful is the decay law[258]

$$\frac{dA}{dt} = kA^n \tag{8.2}$$

where A is activity, k and n constants that depend on process conditions and mechanisms. Equation (8.2) integrates to

$$A = A_0(1 + bt)^m \tag{8.3}$$

where $m = 1/(1 - n)$ and $b = (1 - n)k/A_0^{1-n}$.

For large enough process times $bt \gg 1$ and

$$\log A = m \log b + m \log t \tag{8.4}$$

so that plots similar to that in Fig. 8.3 are found.

Equation (8.4) may be used to predict activity for up to one year, providing accurate data are available in the region of 100–1000 hr. More extensive mathematical modelling is also possible, using rate equations,

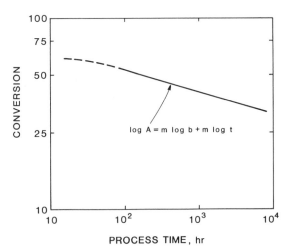

Figure 8.3. Activity versus process time.

deactivation functions similar to equation (8.2), and computer programs simulating particle-reactor behavior. Process designers profit since computer predictions are rapid and inexpensive. The key is fundamental understanding of both reaction and deactivation mechanisms and kinetics. Unfortunately, few processes are that advanced.

Deactivation regimes such as those in Fig. 8.2 are corrected in a number of ways. First, harmful process conditions are avoided or changed. Decreasing (or increasing in some cases) the temperature, increasing hydrogen pressure, etc. are often sufficient to lower deactivation rates. This is often an important consideration in designing process conditions. Another approach is to maintain constant conversion by increasing temperature gradually as the catalyst decays. This is limited by the sensitivity of the process equipment to high temperatures, capacities of furnaces and heat exchanges, and possible side reactions.

Second, poisons in the feed may be removed with up-stream processing units or guard chambers. Catalytic reforming catalysts are poisoned by sulfur compounds in naphthas, so the feed is hydrotreated for desulfurization. Sulfur in the natural gas feedstocks for steam reforming in ammonia production is removed with zinc oxide guard beds. A clever and innovative solution to lead poisoning of platinum–alumina automobile exhaust catalysts is to formulate pellets with subsurface shells of platinum as shown in Fig. 6.16, the outer alumina surface acting as a guard for the platinum.

Third, in processes designed for equilibrium conversion, extra large catalyst beds may be used. The reaction occurs in a narrow zone at the front of the bed. As the catalyst deactivates, this zone moves along the reactor until slippage occurs. Catalyst charges are large enough to allow operation for specified periods between shut-downs. Strongly adsorbing poisons result in the best performance, as experienced with adiabatic processes such as low temperature shift and methanation. Figure 8.4 shows temperature profiles for this type of process, indicating well-defined poisoned, reaction, and clean zones, and differences between different types of deactivation.[260]

Next, reactors may be designed to accommodate rapid deactivation-regeneration cycles. The best example is catalytic cracking where coke deposition is so fast that decay takes only minutes.[261] Since, in this case, the feed is the precursor to coke, treatment or guard chambers are not practical. The only solution is to use fluidized beds, which provide reaction and regeneration in a continuous cycle. Other cases are slurry and moving bed reactors, when deactivation is not so rapid.

Last, but perhaps most important, is catalyst modification. Other solutions should be attempted only when this avenue for catalyst control is exhausted. The most common methods are as follows:

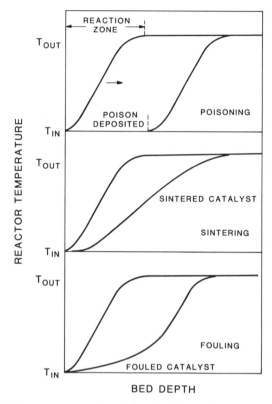

Figure 8.4. Temperature profiles during deactivation in adiabatic reactors.

1. Choice of an alternate but perhaps less active catalyst. It may be worthwhile from an economic viewpoint to substitute either a different active component or another support in order to achieve greater lifetimes, although at the expense of activity.

2. Optimization of the the active component–support combinations. A different support may be used to increase interactions, thereby reducing sintering. Supports with less acidity may be available in cases of excessive coke formation.

3. Addition of promoters to resist deactivation. For example, greater support stability with alumina is achieved with small amounts of added silica or zirconia, sintering and coking of platinum is reduced by adding rhenium, and acid sites neutralized with potassium.

4. Addition of promoters to remove deactivating agents. One example is steam reforming, in which carbon is removed, as it forms, by reaction with steam, catalyzed by alkali additives.

5. Addition of promoters to neutralize poisons. Sulfur poisoning of nickel is reduced in the presence of copper chromite, since copper and chromium ions preferentially form sulfides. Another example is heavy metals poisoning of cracking catalyst, in which iron, nickel, and vanadium are alloyed with antimony added to the feed and deposited on the catalyst.

The proper choice of which approach to follow, catalyst or process modification, requires a detailed knowledge of the cause of the deactivation.

8.3. CAUSE OF DEACTIVATION

Reasons for catalyst deactivation are listed in Table 8.1. It is impossible to completely separate mechanical, thermal, chemical factors. Obviously, temperature influences the sensitivity to poisons as well as to sintering. Other thermal effects may be changed by chemical interaction with the reactants. Similarly, both thermal and chemical forces can weaken a particle to cause mechanical failure. Nevertheless, it is convenient to consider them as distinct causes for deactivation, affecting either the surface concentration of the active sites or the surface area. Sintering, poisoning, and coking are the most important, yet others should also be taken into account during diagnosis and analysis.

8.3.1. Particle Failure

Catalyst particles, if properly selected and installed according to specifications, should have sufficient strength to resist failure due to fracture. However, crushing and attrition tests are run on fresh catalysts. Changes during process operations result in gradual deterioration of mechanical properties, perhaps unevenly, through the bed. Consequences of this are

TABLE 8.1. Causes for Catalyst Deactivation

Type	Cause	Results
Mechanical	Particle failure	Bed channeling, plugging
	Fouling	Loss of surface
Thermal	Component volatization	Loss of component
	Phase changes	Loss of surfaces
	Compound formation	Loss of component and surface
	Sintering	Loss of surface
Chemical	Poison adsorption	Loss of active sites
	Coking	Loss of surface, plugging

plugging, channeling, pressure drop increase, and irregular bed performance. This in turn leads to hot spots, with associated thermal and coking effects. Thus particle failure initiates more serious deactivation by other mechanisms.

Loss of mechanical strength is unlikely, unless assisted by thermal and chemical effects. For example, an increase in the pressure of steam reforming units produced a gradual loss of volatile silicates. Since pellets were cemented together with calcium silicate type binders, silicate loss led to weakening of the catalyst rings. The corrective action was a change in the method of preparation and formulation.[49] Another example is found with alumina-based pellets. Sulfate impurities that are not removed during preparation react with aluminum under regeneration conditions to form aluminum sulfate, weakening the pellet. This also occurs when heavy metals deposit during residual oil desulfurization.[36] Regeneration in air at elevated temperatures produces vanadium oxides which act as a flux to change particle binding. Particle failure is usually obvious, but the contributing factors may not be.

8.3.2. Fouling

In this sense, fouling means deposition of reactor debris on the particles. Scale, rust, and other corrosion products are all possibilities, in addition to chemical components from up-stream units. Particles removed from reactors often have red–brown iron oxide crusts on the outside. Calcium compounds are also found. The most severe cases occur in processing coal and coal-derived liquids, which contain large amounts of inorganic mineral matter.

At best, these materials clog the outside of particles, plugging pores and blocking active surfaces. At the worst, particles are cemented together, resulting in loss of void space within the reactor and complicating catalyst removal. Once again, the analytical procedures discussed in Chapter 7 are helpful to indicate the presence of this type of deactivation. There is no cure, the catalyst can only be replaced, and care exercised to avoid future contamination by the use of screens and guards.

8.3.3. Component Volatization

Elevated temperatures can result in loss of active components or promoters through vaporization. Loss of activity and promoter function follows, with added complications from corrosive deposition on downsteam plant equipment. Events of this type are easily detected in laboratory operations, so we must assume that, in process use, they are only found with unexpected upsets or on a long-term basis.

An example is found in hydrotreating with molybdenum-containing catalysts. During regeneration, coke is burned off with air. Although precautions are taken to avoid reactor hot spots, these sometimes occur. Molybdena volatizes above 800°C, so that yellow crystals deposit downstream, and the catalyst particles turn white. Activity is irreversibly lost. Another case is found with nickel methanation catalysts. If the catalyst bed cools below 150°C in the presence of carbon monoxide-containing gas, nickel carbonyl vaporizes from the particle, creating a very toxic hazard as well as loss of nickel.

Long-term volatization is also found in steam reforming of naphtha. Coke formation from heavier hydrocarbons is controlled with potassium, which promotes carbon–steam reactions. In the presence of steam, however, potassium slowly forms KOH which volatilizes, resulting in accelerated coke formation. The solution is an ingenious example of the catalyst designers expertise. Less than 10% kalsilite (K_2O, Al_2O_3, SiO_2) is included in the catalyst formulation. In the presence of steam and CO_2, kalsilite slowly decomposes to K_2CO_3 and KOH, always providing enough alkali to remove carbon, even though it eventually vaporizes.[49] Lifetimes of 4–5 years have been achieved with these catalysts.

8.3.4. Phase Change

All components in the catalyst must be maintained in their most active state. Phase changes are thermal phenomena and are distinguished from compound formation. High surface area aluminas, for example, change to low area phases when heated. Alumina-based catalysts are often regenerated by burning the carbon. Over a period of time, the surface area declines and activity drops, due to thermal sintering during regeneration. This is one reason why the activity does not return to initial conditions, as shown in Fig. 8.2. This phase change is controlled by adding silica as a promoter.

Segregation of components is also considered a phase change. Highly dispersed alloys are notorious in their tendency to form nonhomogeneous crystallites, with more volatile components diffusing to the surfaces. This is found many times with copper–nickel alloys, confusing much of the early work in metal catalysts.[262] There is also evidence that rhenium and iridium promoters of platinum in catalytic reforming catalysts undergo some degree of segregation upon regeneration.[11]

8.3.5. Compound Formation

Compounds form between components and the reactive atmospheres, rendering the catalyst less active. Elevated temperatures accelerate the reaction of nickel with common supports to produce nickel aluminates and silicates. Steam forms oxides from cobalt and iron, nitrogen results in nitrides, and carbon-containing atmospheres give carbides. These are not surface phenomena, similar to poisoning, but bulk compound formation, easily identified with x-ray diffraction.[263]

8.3.6. Sintering

Sintering is a well-known phenomenon in metallurgy and ceramic science. Much of what we know about this topic comes from these disciplines.[264] However, special features apply to catalysis because of the extremely small crystallite sizes, porous supports, reactive atmospheres, and relatively lower temperatures. As discussed in Chapter 1, surface and bulk atoms become mobile at temperatures of about one-third and one-half of the melting point, respectively. Since supports are high-melting ceramic oxides and active components are low-melting, dispersed metals, oxides, or sulfides, different sintering mechanisms apply to each.

Progressive steps in sintering of a support is shown in Fig. 8.5. Small gains (50–500 μm) with high internal mesoporosity are compressed together during formulation. Grain boundary flow and necking is part of the process which imparts strength to the particle. Sintering has already commenced. At this point, the particle is better described in terms of porosity. Diffusion of the material occurs first across bridges between small pores, where surface tension forces are highest. Initially, small pores collapse leading to decreased

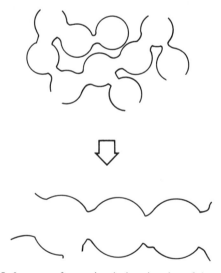

Figure 8.5. Increase of pore size during sintering of the support.

surface area. A typical example is shown in Fig. 8.6 for a naphtha steam reforming catalyst used in the production of substitute natural gas.

The consequences of pore collapse in the support are twofold. First, crystallites of active components are encapsulated in the smallest pores and become inaccessible. Second, the dispersing function of the support is diminished. Supported crystallites move closer together, accelerating loss of active surface as they sinter.

Good catalyst design should anticipate these possibilities. Calcination is carried out at temperatures higher than those encountered in processing. Supports are selected and promoted to impart stability within the desired range. Although supports are least likely to respond to thermal deactivation, many do, testifying to the fact that, during many months of usage, thermodynamics asserts itself. Unexpected events occur as unforeseen interactions appear.

Most theoretical and experimental research on sintering has centered around highly dispersed metals, such as platinum on alumina. This is understandable since these systems are close to ideal, are easily studied, and suffer the greatest deactivation when sintered.[265] Figure 8.7 illustrates two accepted mechanisms for crystallite growth. The first is crystallite migration. Small crystallites have a large fraction of surface atoms, which become mobile at lower temperatures than the bulk. A shape displacement

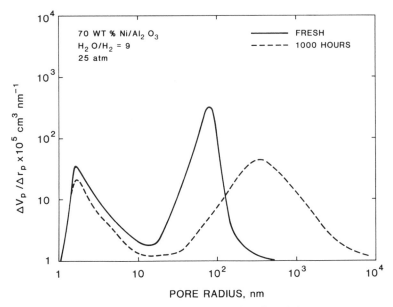

Figure 8.6. Loss of small pores during sintering of the support.

CRYSTALLITE   MIGRATION   INTERPARTICLE TRANSFER

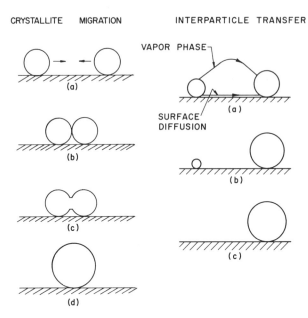

Figure 8.7. Mechanisms for the sintering of dispersed metals.

results which leads to a statistical number of movements. A type of surface Brownian motion, this mobility produces collisions and consequential coalescence. Migration rates may control the kinetics, but these depend on a number of parameters including temperature, crystallite size, surface wetting, and support substrate. For platinum–alumina catalysts, the limiting crystallite diameters appears to be about 5 nm. Figure 8.8 shows crystallite size distributions during sintering of 40 wt % Ni/SiO_2. Progression of smaller crystallites to larger sizes is clearly seen.[266]

The second mechanism is interparticle transport of atoms from small to large crystallites. The driving force is the larger free energy of smaller crystallites, which increases the vapor pressure and evaporation, with condensation occurring on larger crystallites. Distribution changes for this process are shown in Fig. 8.9, where small crystallites disappear as larger ones grow.[267]

Transport may occur over the surface or through the atmosphere, with the latter most likely to involve molecular intermediates. This is true with platinum–alumina, for example, where oxidizing atmospheres during regeneration produce volatile PtO_2 molecules. In nickel-silica catalysts exposed to carbon monoxide, nickel carbonyl serves the same purpose. For surface transport, atomic migration is favored, but depends on the substrate composition.

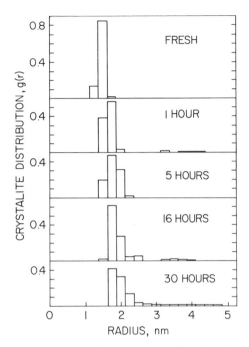

Figure 8.8. Crystallite size distribution changes for crystallite migration of nickel on silica at 500°C.[266]

For catalyst designers, these mechanisms point the way to possible corrective modification. First, promoters may be used to prevent collisions between migrating crystallites. Figure 8.10 shows an example of this in a Cu–ZnO–Al₂O₃ catalyst used in low-temperature water–gas shift processes. Second, intercrystallite transport can be restricted by promoters acting as preferential adsorbents. In Fig. 8.10, copper crystallites are protected from migration by zinc oxide, but sinter in the presence of chlorine. However, high surface area zinc oxide preferentially adsorbs chlorine, further protecting the copper crystallites.

Third, both crystallite and atomic migration may be retarded by modification of the substrate. Alumina surfaces have cationic vacancies which encourage coordination with transition ions. As sites for nickel adsorption, these vacancies provide a mechanism whereby metal atoms (and parts of crystallites) "hop" over the surface. By adding magnesia to form surface MgAl₂O₄, these vacancies are filled and migration inhibited. In fact, there is some support for the model that nickel oxide–alumina preparations result in nickel oxide dispersed on nickel aluminate patches (see Fig. 6.13), upon which nickel crystallites resist sintering. Data on wetting

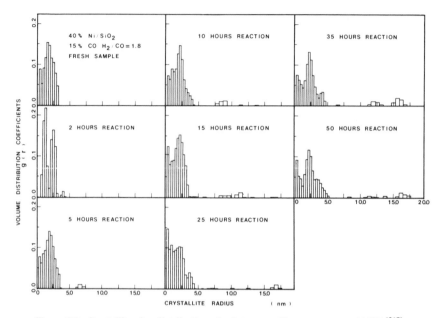

Figure 8.9. Crystallite size distributions for intercrystallite transport at 225°C.[267]

angles of metals and oxides on substrates are very limited, but such information is necessary in order to aid in these designs.

Both mechanisms predict surface decay curves of the form[265]

$$-\frac{ds}{dt} = ks^n \qquad (8.5)$$

with values of n from 2 to 4 for crystallite migration and from 5 to 8 in intercrystallite transport. Actual values vary considerably owing to the effect

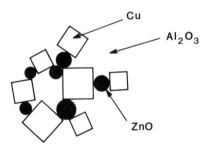

Figure 8.10. Structure of Cu–ZnO–Al$_2$O$_3$ water gas shift catalyst.

TABLE 8.2. Deactivation with Successive Regeneration
0.6% Pt/Al_2O_3—Naphtha Reforming[a]

State	Hydrogen adsorbed $[cm^3(gcat)^{-1}]$
Fresh	0.242
Coked—1 day (1% C)	0.054
Regenerated	0.191
Coked—1 day (1% C)	0.057
Regenerated	0.134
Coked—5 days (2.5% C)	0.033
Regenerated	0.097

[a] Reference 277.

of support curvature in porous substrates and to surface inaccessibilities
that develop when crystallites grow inside small pores.[268] Nevertheless,
equation (8.5) is a useful way to follow sintering changes, providing crystallite size and not surface area is measured.

Loss of activity as the surface declines is the main result of sintering.
This loss may be greater than indicated by size increase alone if surface
inaccessibility also increases.[269] However, another consequence is change
in selectivity. Chapter 3 gives instances where structure demanding reactions
show crystallite size effects. For parallel reactions occurring on different
crystallite sites, large changes in selectivity may result. Consider the case
given in Table 8.2 for regenerated catalytic reforming platinum–alumina
catalyst.[277] These data suggest that burning in oxygen during regeneration
induces crystal growth. Redispersion or splitting of the platinum crystallites
follows chlorine treatment, but the activity is steadily declining. A more
detailed examination shows some interesting features. Table 8.3 gives
changes in the yield as the crystallite size changes.

TABLE 8.3. Change of Selectivity with Crystallite Size n-Heptane Reforming,
0.3% Pt/Al_2O_3, 780°C[a]

Surface area $[m^2g\,(Pt)^{-1}]$	d_c (nm)	Isomerization	Percent yield Dehydrocyclization	Hydrocracking
233	1.0	9.0	37.4	50.6
202	1.2	10.6	32.8	53.1
72	3.3	14.2	26.8	54.4
32	7.3	21.7	21.6	49.7
15	15.8	24.3	17.7	48.2

[a] Reference 277.

Aromatic production from dehydrocyclization decreases with larger crystallite diameter, d_c, whereas isomerization increases and hydrocracking remains approximately the same. Thus, selectivity to the high-octane product decreases with sintering.

In conclusion, sintering is primarily a thermal process of pore collapse and crystallite growth, although chemical interactions with reactants may also play secondary roles. It is normally irreversible but may be effectively controlled with catalyst promoters.

8.3.7. Poisoning

Poisoning is a chemical effect, although temperature may influence sensitivity. A poison is any agent that reacts permanently with an active site. Exceptions are coking and catalyst impurities. Coking has separate characteristics and is considered as a distinct type of deactivation. The effect of impurities in the catalyst is a thermal phenomenon and has been discussed earlier. Poisoning may be reversible or irreversible, regenerable or not, depending on the type of poison, catalyst, and service. Table 8.4 shows reaction schemes which involve poisons.

Most poisons are type (1), i.e., independent compounds present in the feed, perhaps in minute quantities, that deactivate the site with a mechanism different from the main reaction. Examples are also found of types (2) and (3), where either parallel or series reactions generate side products that poison the sites. These mechanisms may also be classified as examples of kinetic inhibition but are considered poisoning if adsorption on the site is irreversible. In situations where multiple sites are involved (for example, dual-functional catalytic reforming), poisoning patterns become more complex.

Chemically, a poison is any compound resulting in strong adsorption. In practice, many different types are encountered. Examples are listed in Table 8.5. Since poisoning is an act of adsorption, two features are important.

TABLE 8.4. Origin of Catalyst Poisoning[a]

Type of reaction
(1) R → P
X → S-X
(2) R → P₁
R → P₂ → X → S-X
(3) R → P₁ → P₂ → X → S-X

[a] R, reactant; P, product; X, poison; S-X, poisoned site.

TABLE 8.5. Types of Poisoning Agents

Type of poison	Origin	Example
Simple ions:		
Cl and other halides	Boiler water	Low temperature shift
Simple molecules:		
S–	Petroleum	Hydrogenation
N–	Petroleum	Acid catalysts
P–	Petroleum	Hydrogenation
As–	Petroleum	Hydrotreating
CO_2	Synthesis gas	Ammonia synthesis
Complex molecules:		
Ni, Fe, V phorphorins	Petroleum	Catalytic cracking
Other metals	Compressor oil	Hydrocracking
Pb compounds	Additives	Hydrotreating
		Automobile exhaust

These are (1) site heterogeneity and (2) diffusion. Adsorption sites may not have uniform attraction for poison molecules, in which cases curves such as those in Fig. 8.11 are encountered.

Curve (1) is found with uniform sites. Decay of activity with poison concentration is linear and follows the form:

$$R = R_0(1 - \alpha) \tag{8.6}$$

where α is the fraction of surface poisoned. In curve (2), less active, and in (3), more active, sites are poisoned first. Curve (4) is an extreme case

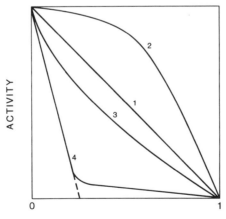

FRACTION OF SURFACE POISONED

Figure 8.11. Loss of activity for different types of poisoning. 1, Uniform; 2, least active sites; 3, most active sites; 4, dual sites.

where there are two types of sites, with the most active eliminated first. An example of this last case is seen in Fig. 8.12. The reaction is cyclopropane isomerization and hydrogenolysis over Ni/SiO_2-Al_2O_3.[61] Poisoning with H_2S decreases both activities at first, but when the temperature is raised only isomerization remains. These data suggest two different sites for these reactions.

The curves in Fig. 8.11, are found with poisoning titration curves discussed in Chapter 7. Interpretation is ambiguous, however, since other mechanisms give similar results. For example, the linear curve (1) is found in cases of strong adsorption. The poison saturates a zone in the front of the bed, which moves uniformly down the reactor, so that α becomes the fraction of unpoisoned bed. Curves similar to (3) originate when an ensemble of n atoms is necessary for the reaction. If the poison has a site stoichiometry of n, then equation (8.6) becomes

$$R = R_0(1 - \alpha)^n \tag{8.7}$$

These complications confuse laboratory interpretations. With process reactors, however, beds are so large that preferential poisoning zones usually occur. For example, methanation reactors are operated adiabatically so that

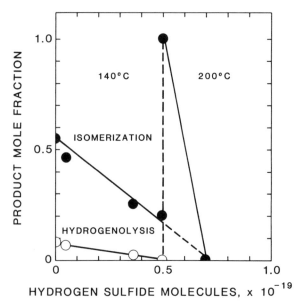

Figure 8.12. Change of selectivity patterns in cyclopropane conversion with hydrogen sulfide poisoning of a nickel catalyst.[61]

the temperature profile becomes a measure of the reaction zone. In Fig. 8.13 we see the progression of a temperature profile down the reactor as the nickel catalyst is poisoned with sulfur. Notice that in case (a) the zone shape remains approximately constant, indicating saturation poisoning behind the front, none ahead of it. Uniform poisoning results in case (b), showing that the amount of catalyst for complete conversion (the reaction zone width) increases or activity decreases in a regular manner. These types of temperature profiles may be used for diagnosis of deactivation mechanisms, especially if combined with mathematical modeling.

When the reaction is pore diffusion controlled (low effectiveness factor, high Thiele modulus), poisoning shows unexpected results. It is necessary to differentiate between two extremes, with and without diffusional resistance to poisoning. For the case of a poison without diffusion problems, analysis of particle kinetics gives the results of Fig. 8.14. Low values of the reactant Thiele modulus, ϕ_R, result in a linear deactivation curve.[32] However, if ϕ_R is large and the process strongly limited by pore diffusion, then the poison has less effect on the initial deactivation rate. Physically this means that since the reaction occurs mostly in the outer layers of the particle, poisons have less effect as they penetrate deeper. The inner part of the particle acts as a sink for the poison. This leads to the interesting conclusion that larger particles deactivate more slowly, even though activity is less. Whether this helps a process situation is entirely a question of

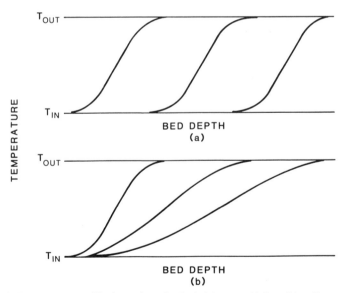

Figure 8.13. Temperature profiles in methanation beds (a) zone sulfiding, (b) uniform sulfiding.

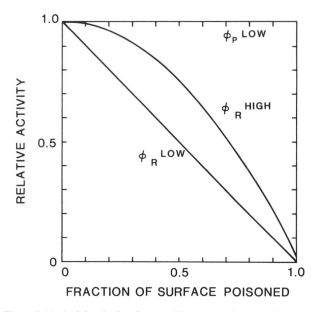

Figure 8.14. Activity decline for nondiffusional resistance of poison.

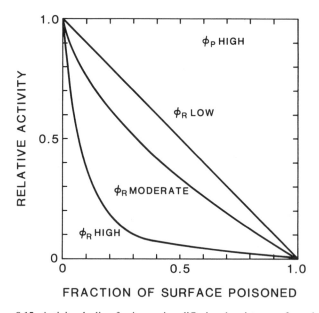

Figure 8.15. Activity decline for increasing diffusional resistance of a poison.

economics. Figure 8.15 shows activity decline when the poison encounters diffusion resistance. Once again, if the reaction has a low value of ϕ_R then decline is uniform. However, with strong diffusion limitations, the reaction concentrates in the outer layers of the particle. Only a small amount of poison is needed to give substantial decreases in activity.

Thus, in the diagnosis and analysis of poisoning deactivation, site heterogeneity, stoichiometry, and diffusion play important roles. Each situation must be considered on its own merits, with all these factors taken into consideration. Techniques for mathematical modeling of poisoning phenomena have advanced to the point where they are useful tools in this endeavor.[259]

We now examine in more detail the poisoning of catalytic materials, maintaining the convenient classification of metals, semiconductors, and insulating acids.

8.3.7.1. Poisoning of Metals

Catalytic metals have d orbitals available for adsorption. This is the key both to activity and susceptibility to poisons. Metal poisons fall into three classes: (1) nonmetallic ions, (2) metallic ions, and (3) unsaturated molecules.[273]

8.3.7.1a. Nonmetallic Ions (s and p Orbitals). Most commonly encountered are those in group Vb (N, P, As, and Sb) and group VIb (O, S, Se, Te). The degree of toxicity depends on the availability of empty valence orbitals or unshared electron pairs. If these are not present, then the ion is nontoxic. Examples are given in Table 8.6. In certain environments, ions are toxic; in others they are nontoxic.

8.3.7.1b. Metallic Ions (d Orbitals). Ions of metals with occupied d orbitals possess electrons that bond with empty orbitals of catalytic metals. In applying this concept to platinum, metal ions were found to be toxic or nontoxic as shown in Table 8.7.

8.3.7.1c. Unsaturated Compounds. Metals have a strong preference to adsorb unsaturated compounds such as CO or C_2H_4. If retained irreversibly in the molecular form, this is the same as poisoning. However, if decomposition or dissociation occurs, it is more properly treated as part of a coking mechanism.

Reversibility of these poisons depends on process conditions. Sulfur-poisoning of nickel catalysts, for example, is irreversible at lower temperatures. Methanation catalysts beds cannot be regenerated even with

TABLE 8.6. Toxicity of Nonmetallic Ions to Metals[a]

Group	Element	Toxic ion	Nontoxic ion
Vb	N	H:N:H (with H below)	$\begin{bmatrix} H \\ \mid \\ H-N-H \\ \mid \\ H \end{bmatrix}^{-}$
	P	H:P:H (with H below)	$\begin{bmatrix} O \\ \mid \\ O-P-O \\ \mid \\ O \end{bmatrix}^{3-}$
	As	H—As—H (with H below)	$\begin{bmatrix} O \\ \mid \\ O-As-O \\ \mid \\ O \end{bmatrix}^{3-}$
Vlb	S	H—S—H	$\begin{bmatrix} O \\ \mid \\ O-S-O \\ \mid \\ O \end{bmatrix}^{2-}$
		R—S—R'	$\begin{bmatrix} R'' \\ \mid \\ R-S-R' \\ \mid \\ O \end{bmatrix}^{2-}$

[a] Reference 273.

hydrogen treatment. Surface sulfur is held much more strongly than in the bulk, so that sulfides may reduce yet retain sulfur-poisoned surfaces.[274] At higher temperatures, however, removal of sulfur by hydrogen and steam is much faster, so that sulfur resistance in steam reforming is much higher.[49]

TABLE 8.7. Toxicity of Metal Ions to Platinum

Toxic ions	Nontoxic ions
Zn^{2+}	Li^+ Be^{2+}
Cd^{2+} In^{3+}	Na^+ Mg^{2+} Al^{3+}
Hg^{2+} Sn^{2+}	K^+ Ca^{2+}
Hg^+	Rb^+ Sr^{2+} Zr^{4+}
Tl^+ Pb^{2+}	Cs^+ Ba^{2+} La^{3+}
	Ce^{3+}
$Cu^+Cu^{2+}Ag^+Au^+$	Th^{4+}
$Mn^{2+}Fe^{2+}Co^{2+}Ni^{2+}$	Cr^{2+} Cr^{3+}

Furthermore, the surface can be cleaned with sulfur-free purge gas. Regenerative processes for chlorine- and sulfur-poisoned low temperatures shift catalysis are available, but there is some doubt as to desirability.

In general, the best approach to poisons of this type is to pretreat the feed. This is done in three ways: (1) chemical treating, which is expensive and may introduce other contaminants; (2) catalytic processing, which is effective for poisons with organic origins; and (3) guard chambers. In guard chambers, a cheaper adsorbent precedes the main reactor. Sometimes this is a separate unit, such as sulfur-removing zinc oxide beds in natural gas reforming. On other occasions it may be part of the reactor bed and the front section is sacrificed as a scavenger, using the same catalyst or a cheaper substitute. Some very ingeneous guard applications are found in automobile exhaust catalysts, as discussed in previous sections.[255]

8.3.7.2. Oxides

Semiconducting oxide catalysts owe their activity to electron accepting or donating surface sites, with specific surface geometry and favoring redox reactions. Any molecule which adsorbs strongly constitutes a potential poison. Unfortunately, not too much attention has been given to this subject in the literature.[256] For example, poisoning of hydrogenation or hydrogenolysis activity in supported oxides and sulfides has not been considered to any extent. Other types of deactivation take precedence. Other than these general remarks, very little information can be added.

8.3.7.3. Solid Acids

Poisoning of acid sites is straightforward. Basic constituents are required to neutralize acidity. This is found in alkali and alkaline earth compounds and in basic organic molecules. Process poisoning by alkaline and alkaline earths is rare. These materials are added as deliberate promoters to remove acidity but are not normally encountered in process streams in the basic form. One exception is Na^+ ions, encountered in steam used for stripping cracking catalysts and other purposes.

Basic organics are primarily nitrogen-containing molecules that are abundant in petroleum feedstocks. In a typical straight run stock, 25%–35% of the nitrogen compounds are basic. Basic and nonbasic types are shown in Table 8.8.

The susceptibility of solid acids to these poisons correlates well with the basicity of nitrogen compounds, as shown in Fig. 8.16. In practical feeds, a wide range and variation of types are present.

TABLE 8.8. Basic and Nonbasic
Compounds in Petroleum

Type	Compound
Basic	Pyridines
	Quinolines
	Amines
	Indolines
	Hexahydrocarbazoles
Nonbasic	Pyrroles
	Indoles
	Carbazoles

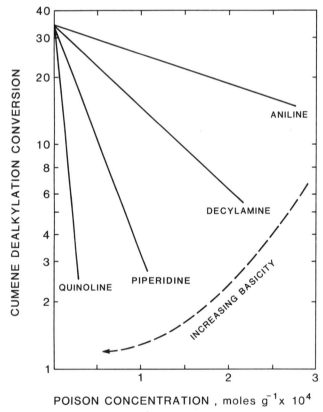

Figure 8.16. Poisoning of acidity by basic nitrogen compounds.

These poisons are removed during regenerative burning but may cause problems by producing NO_x in combustion effluents. It is better to remove them before the process. Hydrotreating removes nitrogen, and this is standard procedure with all catalytic cracking feedstocks. This treatment also removes sulfur and much of the heavy metal contamination.

8.3.7.4. Beneficial Poisons

Poisons generally result in harmful effects such as loss of activity, yet there are circumstances when small amounts of poison are beneficial. An example of this is catalytic reforming to boost the octane number of paraffinic naphthas. Early workers noticed that a few parts per million of sulfur in the feed increased reformate yield rather than decreasing as expected. This "promotion" technique has become standard in reforming technology.[277] The reason for this is seen in studies with pure compounds on nickel–faujasite catalysts.[246] Nickel results in more hydrogenolysis than platinum. This reaction leads to light gases and coke and is undesirable over dual-functional isomerization. With n-hexane as the feed, the fresh catalyst showed poor selectivity to isomerization, as shown in Fig. 8.17. Poisoning with hydrogen sulfide, however, showed a dramatic selectivity reversal. Sulfur poisons the most active hydrogenolysis sites preferentially.

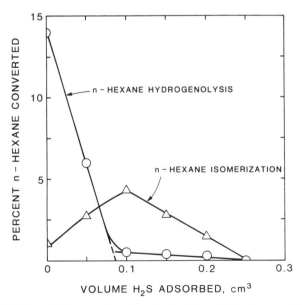

Figure 8.17. Beneficial poisoning of a nickel catalyst with hydrogen sulfide.[246]

These are most likely edge-corner atoms. Additional poisoning destroys all activity eventually. In cases such as this, selective poisoning of undesired activity gives improved yields of desired products.

8.3.8. Coking

"Coke" is a term given to carbonaceous residues on a catalyst surface.[278] Varying in structure from hydrogen-deficient, aromatic-type polymers to graphic carbon, these deposits are found in amounts up to 20 wt % whenever carbon-containing feeds are exposed to catalytic surfaces. All catalysts experience this deactivation to some extent. In extreme cases, the catalyst surface is covered with layers of coke deposit, thereby decreasing the accessible surface areas, active components become encapsulated in carbon, and pores are blocked with heavy buildup. When these effects become significant, the coked catalysts must be replaced or regenerated.

Coke accumulation is the difference between coke deposition and removal. Both occur, although at different rates. Well-designed catalysts provide an economic balance between the two. Deposition from hydrocarbons occurs in two ways, on acid sites and on dehydrogenation sites:

Acid sites: (1) $C_n H_m \rightarrow (CH_x)_y$ (8.8)

Dehydrogenation sites: (2) $C_n H_m \rightarrow yC$ (metals and oxides)

With carbon monoxide and dioxide, two other possibilities occur:

Dissociation sites: (3) $2CO \leftrightarrows C + CO_2$ (8.9)

                        (4) $CO_2 \leftrightarrows C + O_2$

Coke removal involves a number of reactions

                        (5) $C + O_2 \leftrightarrows CO_2$

                        (6) $C + 2H_2 \leftrightarrows CH_4$
                                                                    (8.10)
                        (7) $C + H_2O \leftrightarrows CO + H_2$

                        (8) $C + CO_2 \leftrightarrows 2CO$

where C represents the carbon-containing residue.

Within this framework, coking phenomena are best considered in terms of three common operations: (1) acid coking, for example, cracking on catalysts; (2) dehydrogenation coking, as in catalytic reforming; and (3) dissociative coking, found in steam reforming.

8.3.8.1. Acid Coking

Acid coke forms on silica alumina and zeolite cracking catalysts and on acidic supports. Coke forming tendency is directly related to acidity. There are two major types of carbon structures existing as highly dispersed phases in the pores. Most of the coke is present as pseudographitic or turbostratic, random-layer lattices, similar to graphite, with a composition of $CH_{0.4}$ to $CH_{0.5}$. The remainder consists of poorly organized polynuclear aromatic macromolecules.[279]

There is strong evidence that acid coke originates with aromatic and olefinic hydrocarbons, either initially in the feed or generated as intermediates during the cracking process. Coke buildup correlates well with aromatic/naphthene ratios in gas oils and also with the basicity of polynuclear aromatics, as shown in Fig. 8.18. These molecules readily form ion radicals with acid sites, polymerize with other unsaturates, and then dehydrogenate to aggregates of coke. Olefins likewise play a key role in cracking reactions. They are created through dehydrogenation and act as hydrogen acceptors to form carbonium ions. Unsaturated ions are strongly adsorbed and become increasingly hydrogen deficient, ultimately forming coke via cyclization.

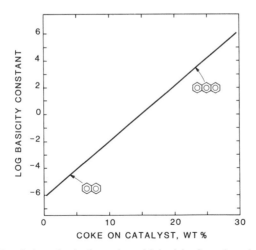

Figure 8.18. Correlation of coke formation with basicity for polynuclear aromatics.

Increased acid site strength and density favor coke formation, which is a fact of life in catalytic cracking and must either be accepted or the catalyst modified to hinder formation or to facilitate removal.

Coke buildup with time is illustrated in Fig. 8.19. The data fit an equation in the form

$$C_c = At^n \tag{8.11}$$

where C_c is the coke formed at process time t, A is a constant that correlates with aromatic content, and n is between 0.3 and 0.5. Equation (8.11) is called the Voorhies equation.[280] The carbon content also correlates with conversion level or severity of operation, as shown in Fig. 8.20, so that combining the two gives plots like these in Fig. 8.21, which obey the general deactivation equation (8.4) with $m = -0.5$. Figure 8.21 also illustrates the benefits derived from adding small amounts (about 5%–10%) of zeolite to amorphous cracking catalyst.[261] The activity is orders of magnitude higher and coke formation greatly reduced. This is a consequence of shape-selectivity discussed earlier, where coke-forming intermediates are restricted by the size of the zeolite cavities.[281]

Another type of coke is identified in catalytic cracking. This is "contamination" coke originating in the dehydrogenation reactions catalyzed with metals deposited by nickel, iron, and vanadium porphorins in the feed. This is discussed in the section on dehydrogenation coke.

We have seen how zeolites reduce the rate of coke formation through control of the size of intermediates. Since the carbon-forming reaction

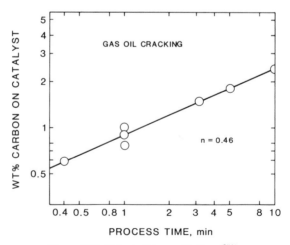

Figure 8.19. Coke build-up with time.[281]

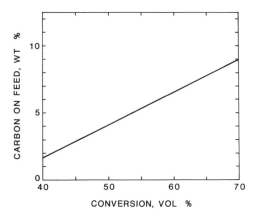

Figure 8.20. Correlation of coke formation with conversion.[281]

parallels the main functions of catalytic cracking, it is difficult to selectively deactivate coking sites. Perhaps when more is known about ensemble effects in acid reactions, it may be possible to control site geometry and discourage formation of larger molecular species without affecting rupture into smaller ones. If acid catalysts are not desired but coking occurs on acidic support sites, then control is possible by substituting nonacidic supports such as magnesia for silica or alumina. If this is not practical, the acid sites may themselves be poisoned. This is usually accomplished with small amounts

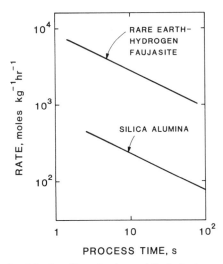

Figure 8.21. Decay of activity for silica alumina and faujasite-type cracking catalyst.[281]

of potassium. Other alkalis and alkaline earths function equally well, but potassium is the most effective.

The other alternative is to remove the coke either at intervals or as it forms. Reactions (8.10) offer pathways, but in many processes, for example, catalytic cracking, the reactions are not compatible with the process. In these situations, the catalyst is regenerated at intervals with mixtures of air and steam via reactions (5), (7), and (8) in (8.10). Regeneration intervals vary with rate of coke buildup and cover a wide range from minutes for catalytic cracking, weeks for catalytic reforming, to months for hydrotreating. The usual procedures are to purge the bed with steam or nitrogen into which air is admitted slowly enough to control exothermic temperature increases.[257]

In catalytic cracking, regeneration is an integral part of the process energy balance. The catalyst is fluidized and passes continuously through reaction and regeneration stages. Cracking is endothermic and the necessary heat is provided by combustion of coke in the regenerator. Coke level and energy requirements are designed to match. Catalyst designers must beware not to invent catalysts that are too coke resistant! Most heat is recovered with complete combustion to carbon dioxide, but this raises regeneration temperatures and causes catalyst sintering. For many years, coke was removed by partial combustion and heat recovered in an afterburner (CO oxidation). This was expensive and less efficient. More sinter-resistant catalysts have been developed, but a better solution is to add minute (ppm) amounts of platinum or other oxidation components to the cracking catalyst. The level is low enough not to promote serious dehydrogenation coking, but high enough to catalyze complete combustion at controllable temperatures.[261] This practice is now becoming widespread.

Regeneration of catalysts in other processes is not so critical since energy balance is not required. Combustion promoters are not necessary. However, regeneration facilities must be provided and the plant shut down for a certain period—both expensive. For less frequent regenerations, it is advantageous to replace the catalyst with a fresh load and regenerate off-site. A considerable industry is growing around these operations.

8.3.8.2. Dehydrogenation Coking

When dehydrogenation catalysts (metals, oxides, or sulfides) are present, then the cascade of reactions leads to carbon in a form somewhat different from that on acidic sites. Figure 8.22 shows the process whereby dehydrogenation and associated hydrogenolysis lead to carbon fragments C_x. The carbon is reactive and exists as a surface carbide-type or as pseudographite. With acid supports, these fragments migrate from the

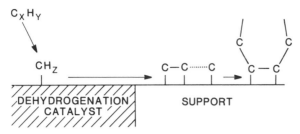

Figure 8.22. Carbon formation on dehydrogenation catalysts.

dehydrogenation surface to fuel the reactions discussed for acid catalysts. Both types of coke are present but in much lesser amounts than for cracking, since acidities are lower and hydrogen from dehydrogenation helps keep the catalyst clean. The carbon-forming tendencies of these catalysts correlate well with hydrogenolysis activity, as expected, since C–C bonds rupture in the presence of hydrogen. Therein lies the best possibility for control, since we have seen many examples of hydrogenolysis suppression through catalyst modification.

The characteristics of dehydrogenation coking vary with the chemistry of the catalytic process. Important features are best considered with three important examples: (1) catalytic reforming, (2) hydrotreating, and (3) metal contamination.

8.3.8.2a. Catalytic Reforming. Catalytic reforming combines two carbon-producing functions: dehydrogenation and acidity. For many years, processes that upgraded octane number with hydrodecyclization, isomerization, and hydrocracking of straight chain paraffins and naphthenes used dispersed platinum on acidified alumina. Much research has been done on carbon formation by these metal crystallites, and a clearer insight is beginning to emerge.[282] Carbon is produced through hydrogenolysis at sites that correlate well with low index positions, such as faces, edges, and corners. Carbon diffuses over the surface as a type of surface carbide and through the bulk as carbon atoms reaching the interface between platinum and support. Some of it progresses to acid sites to begin the polymerization process leading to pseudographite. Thus the metal "feeds" the coke precursor to acid sites and carbon forms in regions around the crystallites. In extreme cases, ribbons of graphitic carbon grow, supporting the platinum crystallite at its tip. Ultimately, these ribbons and patches cover the surface and block pores.[283]

Early workers in catalytic reforming discovered that a small amount of sulfur poisons hydrogenolysis sites and reduces coking. Studies with model components have given a better understanding of this effect, as shown with data in Table 8.9.[246] Initially most of the product comes from hydrogenolysis. A small amount of poison decreases hydrogenolysis but not acid cracking, as measured by the concentration of C_3 hydrocarbons. Finally, all hydrogenolysis is suppressed but with some loss of activity. Yields of isomers, however, increase, as seen in Figure 8.17.

In spite of these improvements, deactivation in catalytic reforming was severe and regeneration was required every few days. This was accomplished with extra "swing" reactors or moving bed designs, both expensive to build and operate. Another solution was to increase pressure from about 15 to 35 atm. Higher hydrogen pressure reduces hydrogenolysis and accelerates hydrogenation of the carbon. Processes then ran three to six months before regeneration. However, aromatic yield from the dehydrogenation of naphthenes and cyclization of n-paraffins decreased and octane number suffered.

A breakthrough came with the discovery of bimetallic reforming catalysts.[11] When platinum is combined with rhenium, the stability improvement is dramatic. Rhenium appears to inhibit both coking and sintering, so that deactivation during use and the frequency of regeneration is decreased. Use of iridium was a further advance, giving not only prolonged life of over ten years but higher activity. Comparisons are shown in Fig. 8.23. Not only was operation easier but plant configuration was simplified with large savings in capital cost. Problems exist with bimetallics, however: which tend to segregate during regeneration, requiring more complicated treatment. But that seems a small price to pay.

TABLE 8.9. H_2S Poisoning of Ni–LiY Effect on Product Distribution n-C_6 Hydrogenolysis and Isomerization[a]

Product	Moles of product Fraction of Ni poisoned with H_2S				
	0	0.05	0.10	0.15	0.20
C_1	56.5	32.8	—	—	—
C_2	—	—	—	—	—
C_3	0.89	2.10	0.64	0.25	0.20
iC_4	0.05	0.23	—	—	—
nC_4	1.63	2.69	—	—	—
iC_5	0.24	0.53	—	—	—
nC_5	2.33	3.43	—	—	—
2MP	0.62	1.66	3.43	2.42	1.01
3MP	0.32	1.42	0.84	0.66	0.49
nC_6	85.5	91.2	95.4	96.7	—

[a] Reference 246.

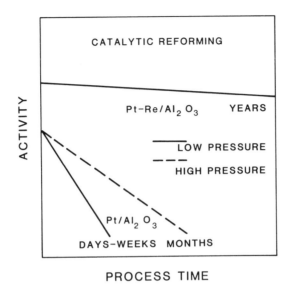

Figure 8.23. Deactivation in catalytic reforming.

8.3.8.2b. Hydrotreating. The principal application of hydrotreating is in desulfurization and denitrogenation of feedstocks for catalytic reformers and crackers. As more efficient use of petroleum is demanded, refiners are processing heavier fractions, so that now it is necessary to hydrotreat a wide range of hydrocarbons from light naphthas through gas oils to vacuum residua. The catalyst is cobalt and nickel-promoted molybdena on alumina, sulfided in use. Many references have been made to this catalyst in previous chapters. Severity of coking is proportional to the molecular weight or boiling range of the feed and to the aromatic content. Whereas for naphthas, coking is moderate, with regeneration necessary only every 6–12 months, deactivation is severe for residua. Levels of up to 20 wt % carbon can build in a matter of weeks.

Acidity of the support, which accelerates coke formation, is poisoned with alkali addition. The active component, promoted molybdenum sulfide, is active for dehydogenation and also has acid sites, so the general behavior discussed for catalytic reforming is the same, except for reduced activity and complications from heavier molecules.

In this system, control of deactivation takes a different course.[36] For heavier fractions, the excessive carbon formation results from asphaltenes. These are large molecules, much more complex than the organic sulfur- and nitrogen-containing compounds that are hydrotreated. The problem was reduced by taking advantage of pore shapes as shown in Fig. 8.24. In

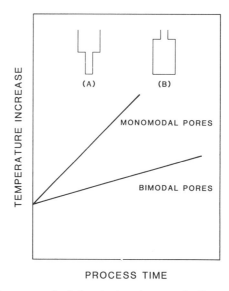

Figure 8.24. Pore shape control of deactivation due to coke formation on hydrotreating catalysts.

(a) small pores are accessed through larger ones, giving a pronounced bimodal pore size distribution. Smaller heteromolecules diffuse into the narrow pores where they desulfurize. Larger ones are restricted to the outside. For (b), "ink bottle" structures with small neck openings allow reactants to pass but exclude coke-making asphaltenes. Both approaches result in significant improvements in catalyst lifetimes.[36]

Since hydrogenolysis of heteroatoms is the desired function it is difficult to reduce coking through selective poisoning. Perhaps more detailed information about the exact structure of the active site will result in better opportunity to selectively control the C-C rupture without affecting C-S and C-N.

As with cracking catalysts, deposition of metal contamination causes problems, and this is discussed further in the next section.

8.3.8.2c. Metal Contamination. Petroleum contains impurity metals in the form of porphorins with the following levels: Fe (0–150 ppm), Ni (0–50 ppm), and V (0–100 ppm). When contacting surfaces such as alumina and silica–alumina, these porphorins adsorb and subsequently decompose to yield a highly dispersed metal. These metals, or corresponding oxides, block surface sites and pores, but the main damage comes from their catalytic activity. First, dehydrogenation activity results in the type of coke initiation discussed in the previous section. Figure 8.25 shows the effects of adding

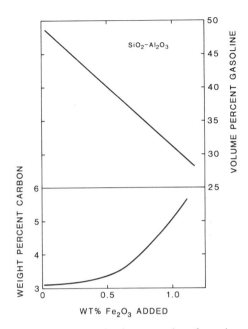

Figure 8.25. The effect of metal contamination on carbon formation with silica alumina catalysts.

this contamination to cracking catalyst. Coke buildup becomes excessive and the dynamics and energy balance of the process are disturbed. Among the three, nickel is the worst, followed by iron and vanadium.

Second, during regeneration these metals oxidize and act as oxidation catalysts, leading to excessive combustion rates and sintering. Especially bad is V_2O_5, not only because it is a strong oxidation agent but also because it melts and forms a flux to accelerate particle degradation.[69]

The best results are achieved by removing the heavy metals from the feed. Chemical or adsorptive treatments to remove the porphorins have been successful, and a large amount of research is now devoted to demetallization. It is interesting that alumina is a good adsorbent, and, since modern cracking units are usually preceded by hydrotreating processes, the alumina-based catalysts in these units act as guards for the cracking catalyst.

Another approach has been deactivation of the metals as they form. Thermal deactivation is a possibility, but this is often destructive to the catalyst. One highly successful solution is to use an additive in the feedstock. Antimony-containing materials, which deposit on the catalyst and effectively passivate the metals, are added to the feed. This has resulted in substantial savings in operations.

As mentioned earlier, metal porphorins also adsorb on hydrotreating catalysts, especially when using heavier residua fractions. The same problems develop, i.e., excessive coke and run-away regeneration. Pore-shape control is practiced, as in Fig. 8.24, for the same reasons. The larger porphorin molecules are excluded to the outer surface of the particles. Ultimately, metal buildup becomes so severe that the catalyst must be replaced. Extraction methods for removing these deposits are available, but as yet have not proven economical. It is interesting that discarded catalysts may become a useful source of nickel and vanadium metal. The best solutions appear to be external guards with cheaper adsorbents, or larger hydrotreating beds, which allow for metal trapping at the top.

8.3.8.3. Dissociative Coking

Dissociative Coking[283] occurs when carbon monoxide dissociates on a catalytic site, as follows:

$$2CO \leftrightarrows C + CO_2 \qquad (8.12)$$

This reaction is found primarily in steam reforming of hydrocarbons, where other coking possibilities also exist. With methane a dehydrogenation reaction

$$CH_4 \leftrightarrows C + 2H_2 \qquad (8.13)$$

occurs. Equilibrium constants for these are shown in Fig. 8.26, together with the third reaction

$$CO + H_2 \leftrightarrows C + H_2O \qquad (8.14)$$

Carbon formation is a major concern in methane and naphtha steam reforming, but Fig. 8.26 shows that under reaction conditions only reaction (8.13) is possible in the temperature range commonly used, 600–750°C. Thermodynamics favors the reverse of reactions (8.12) and (8.14), and carbon is removed, but the catalyst must have sufficient activity to counterbalance coke deposition via reaction (8.13). As discussed earlier, this is accomplished by promotion with potassium. Nevertheless, interesting interactions occur, which may lead to reactor failure. Figure 8.27 illustrates the case for a catalyst deactivated either by sulfur or carbon deposition. Below the equilibrium line, we expect carbon formation. At the inlet of the reactor tube, the temperature dips as the endothermic reaction absorbs heat. If it drops below the equilibrium line, then coke forms and removal rates must

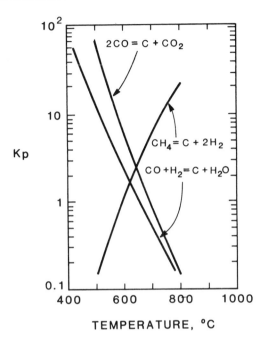

Figure 8.26. Equilibrium constants for coking reactions over nickel catalysts.

be great enough to remove it. Figure 8.27 also shows relative deposition and removal rates for normal catalysts. The danger zone is between the crossover point and the equilibrium line, where both thermodynamics and kinetics favor deposition. Carbon lay-down occurs in this region, which in this example is about one-third of the way down the tube. As the catalyst deactivates in this zone, it loses the ability to react and absorb heat, so the tube heats and a "hot band" develops, leading to tube rupture. In practice, this is avoided by the use of high-activity catalysts in the front part of the tube, thus maintaining sufficient conversion to prevent the hot band condition.[38]

8.4. SOME FINAL COMMENTS

In this chapter we have seen many examples of catalyst modification to counter deactivation. Most of the significant situations in commercial operations have been covered. Those that have not respond to similar solutions. It is hoped that the reader will conclude from the discussion that promotion is not haphazard and has reasonable scientific justification.

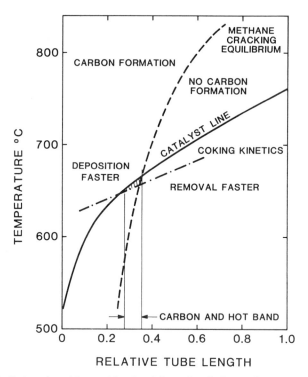

Figure 8.27. Carbon deposition and hot band formation during methane steam reforming.

Similar analyses of unknown situations will no doubt lead to solutions as successful and innovative as those given.

With the end of this chapter we conclude this excursion into the principles of catalyst development. We have learned much about the nature of catalysts, what they do, what they are, and how to select, prepare, test, and preserve them. Guidance has come from the work of many brothers passed before us and inspiration from contemporaries. We have seen how science and technology are pushing back the veil of mystery and alchemy, and yet the magic remains.

As for the future, many exciting discoveries await us, leading ultimately to the day when catalysts are indeed scientifically designed. Catalysis is the key to innovative developments in chemical processing, both for profit and health. The reader is urged to follow the guidelines put forth here but always to find his or her own style. Good luck. May your search bring you as much fun as it has for me.

APPENDIXES

APPENDIX 1
Reference Books and Journals on Catalysis

1. GENERAL CONCEPTS IN CATALYSTS

1.1. Afansiev, V. A., and Zarkov, G. E., *In the Realm of Catalysis*, Mir Publishers, Moscow (1979).

1.2. Ashmore, P. G., *Catalysis and Inhibition of Chemical Reactions*, Butterworth and Co., London (1963).

1.3. Anderson, J. R., *Structure of Metallic Catalysts*, Academic Press, New York (1975).

1.4. Balandin, A. A. (Editor), *Catalysis and Chemical Kinetics*, Academic Press, New York (1964).

1.5. Bartok, M. (Editor), *Stereochemistry of Heterogeneous Metal Catalysts*, John Wiley, New York (1985).

1.6. Basalo, F., and Burwell Jr., R. L. (Editors), *Catalysis: Progress in Research*, Plenum Press, New York (1973).

1.7. Bell, A. T., and Hegedus, L. L. (Editors), *Catalysis Under Transient Conditions*, ACS Symposium Series No. 178, American Chemical Society, Washington, D.C. (1982).

1.8. Bell, R. P., *Acid–Base Catalysis*, The Clarendon Press, Oxford (1941).

1.9. Berkman, S., Morrell, J. C., and Egloff, G., *Catalysis, Inorganic and Organic*, Reinhold Publishing Corp., New York (1940).

1.10. Boldyrev, V. V., Bulens, M., and Delmon, B., *The Control of the Reactivity of Solids*, Elsevier, Amsterdam (1979).

1.11. Bond, G. C., *Catalysis by Metals*, Academic Press, New York (1962).

1.12. Bond, G. C., *The Principles of Catalysis–Monograph for Teachers # 7*, Royal Institute of Chemistry, London (1968).

1.13. Bond, G. C., *Heterogeneous Catalysis, Principles and Applications*, Clarenden Press, Oxford (1974).

1.14. Bosnich, B., *Asymmetric Catalysis*, Martinus Nijhoff Publishers, The Hague (1986).

1.15. Bornelli, Delmon, B., and Derouane, E. (Editors), *Surface Properties and Catalysis by Non-Metals*, D. Reidel Publishers, Dordrecht (1982).

1.16. Burton, J. J., and Garten, R. L. (Editors), *Advanced Materials in Catalysis*, Academic Press, New York (1977).

1.17. Collier, C. H. (Editor), *Catalysis in Practice*, Reinhold, New York (1957).

1.18. Davis, B. H., and Hettinger, W. P. (Editors), *Heterogeneous Catalysis: Selected American Histories*, ACS Symposium Series No. 222, American Chemical Society, Washington, D.C. (1983).

1.19. Dehmlow, E. V., and Dehmlow, S. S., *Phase Transfer Catalysis*, Verlag Chemie, Weinheim (1983).

1.20. Drauglis, E., and Jaffee, R. I. (Editors), *The Physical Basis for Heterogeneous Catalysis*, Plenum Press, New York (1975).

1.21. Emmett, P. H., *Catalysis Then and Now*, Franklin Publishing Co., Englewood (1965).

1.22. Ellis, C., *Hydrogenation of Organic Substances, Including Fats and Fuels*, Van Nostrand, New York (1930).

1.23. Fendler, J. H., and Fendler, E. J., *Catalysis in Micellar and Macromolecular Systems*, Academic Press, New York (1975).

1.24. Falk, K. G., *Catalytic Action*, Chemical Catalog Co. Inc., New York (1922).

1.25. Gates, B. C., Katzer, J. R., and Schuit, G. C. A., *Chemistry of Catalytic Processes*, McGraw-Hill, New York (1979).

1.26. Gates, B. C., Guczi, L., and Knozinger, H. (Editors), *Metal Clusters in Catalysis, Studies in Surface Science and Catalysis, 29*, Elsevier, New York (1986).

1.27. Germain, J. E., *Catalytic Conversion of Hydrocarbons*, Academic Press, New York (1969).

1.28. Golodets, G. I., *Heterogeneous Catalytic Reactions Involving Molecular Oxygen*, Elsevier, New York (1983).

1.29. Green, S. J., *Industrial Catalysis*, Ernest Benn Limited, London (1978).

1.30. Griffith, R. H., and Marsch, J. D. F., *Contact Catalysis*, Oxford University Press, London (1957).

1.31. Hartley, F. R., *Supported Metal Complexes*, D. Reidel Publishing Co., Dordrecht (1985).

1.32. Hata, K., *New Hydrogenation Catalysts, Urushibara Catalysts*, John Wiley, New York (1971).

1.33. Heinemann, H., and Somorjai, G. A. (Editors), *Catalysis and Surface Science*, Marcel Dekker, Inc., New York (1985).

1.34. Hilditch, T. P., and Hall, C. C., *Catalytic Processes in Applied Chemistry*, Van Nostrand Company, New York (1937).

1.35. Ipatieff, V. N., *Catalytic Reactions at High Pressures and Temperatures*, MacMillan, New York (1937).

1.36. Jacobs, P. A., *Carboniogenic Acitivity of Zeolites*, Elsevier, New York (1977).

1.37. Jennings, J. R. (Editor), *Selected Developments in Catalysis*, Blackwell Scientific Publishing Co., London (1985).

1.38. King, D. A., and Woodruff, D. P. (Editors), *The Chemical Physics of Solid Surfaces and Heterogeneous Catalysis*, Vol. 4, *Fundamental Studies in Heterogeneous Catalysis*, Elsevier, New York (1982).

1.39. Krylov, O. V., *Catalysis by Non-Metals*, Academic Press, New York (1970).

1.40. Mukhlyonov, I. P., Dobkina, E. I., Deryuzhkina, V. I., and Soroko, V. E., *Catalyst Technology*, Mir, Moscow (1976).

1.41. Nyrop, J. E., *The Catalytic Action of Surfaces*, Levin and Manksgaord, Copenhagen (1937).

1.42. Ozaki, A., *Isotopic Studies of Heterogeneous Catalysis*, Academic Press, New York (1977).

1.43. Paal, Z., and Menon, P. G., *Hydrogen Effects in Catalysis, Fundamentals and Practical Applications*, Marcel Dekker, New York (1988).

1.44. Pettre, M., *Catalysis and Catalysts*, Dover Publications, New York (1963).

1.45. Pines, H., *The Chemistry of Catalytic Hydrocarbon Conversions*, Academic Press, New York (1981).

1.46. Piszkiewicz, D., *Kinetics of Chemical and Enzymed-Catalyzed Reactions*, Oxford University Press, London (1977).

1.47. Rao, P. T., *Advances in Catalysis, Science and Technology*, Wiley Eastern Ltd., New Delhi (1985).

1.48. Richardson, J. T., *Principles of Catalyst Development*, Plenum Publishing Corp., New York (1989).

1.49. Rideal, E. K., *Concepts in Catalysis*, Academic Press (1968).

1.50. Robertson, A. J. B., *Catalysis of Gas Reactions by Metals*, Springer-Verlag, Berlin (1970).

1.51. Romanowski, W., *Highly Dispersed Metals*, John Wiley, New York (1987).

1.52. Rylander, P. N., *Catalytic Hydrogenation in Organic Synthesis*, Academic Press, New York (1979).

1.53. Sabatier, P., *Catalysis in Organic Chemistry*, Van Nostrand Co., New York (1922).

1.54. Satterfield, C. N., *Heterogeneous Catalysis in Practice*, McGraw-Hill, New York (1980).

1.55. Schwab, G-M., *Catalysis*, Van Nostrand, New York (1937).

1.56. Sinfelt, J. H., *Bimetallic Catalysts, Discoveries, Concepts, and Applications*, John Wiley, New York (1983).

1.57. Shapiro, B. L. (Editor), *Heterogeneous Catalysis*, Texas A & M Press, College Station (1984).

1.58. Shilov, A. E., *Activation of Saturated Hydrocarbons by Transition Metal Complexes*, D. Reidel Co., Dordrecht (1984).

1.59. Starks, C. M., and Liolta, B., *Phase Transfer Catalysis, Principles and Techniques*, Academic Press, New York (1981).

1.60. Stiles, A. V., *Catalyst Supports and Supported Catalysts*, Butterworths Publishing Co., London (1987).

1.61. Stucky, G. D., and Dwyer, F. G., (Editors), *Intrazeolite Chemistry*, ACS Symposium Series No. 128 (1983).

1.62. Szabo, Z. G., and Kallo, K., *Contact Catalysis*, Vols. I and II, Elsevier, New York (1976).

1.63. Tamuru, K., and Ichikawa, M., *Catalysis by Electron Donor Complexes*, John Wiley, New York (1975).

1.64. Tanabe, K., *Solid Acids and Bases*, Academic Press, New York (1970).

1.65. Thomas, J. M., and Thomas, W. J., *Introduction to the Principles of Heterogeneous Catalysis*, Academic Press, New York (1967).

1.66. Thomson, S. J., and Webb, G., *Heterogeneous Catalysis*, John Wiley and Sons, New York (1968).

1.67. Vol'kenstein, F. F., *The Electronic Theory of Catalysis on Semiconductors*, Pergamon, New York (1963).

1.68. Weisser, O., and Landa, S., *Sulphide Catalysts, Their Properties and Applications*, Pergamon, New York (1973).

1.69. Westley, J., *Enzymic Catalysts*, Harper and Row, New York (1969).

1.70. Whyte, Jr., T. E., Dalla Betta, R. A., Derouane, E. G., and Baker, R. T. K. (Editors), *Catalytic Materials: Relationship between Structure and Reactivity*, ACM Symposium Series No. 248, American Chemical Society (1984).

1.71. Yermakov, Y. I., Kuznetsov, B. N., and Zakharov, V. A., *Catalysis by Supported Complexes*, Elsevier, New York (1981).

2. ADSORPTION AND RELATED PHENOMENA

2.1.   Benard, J. (Editor), *Adsorption on Metal Surfaces*, Elsevier, New York (1983).
2.2.   Chattoraj, D. K., and Birdi, K. S., *Adsorption and Gibbs Surface Excess*, Plenum Press, New York (1984).
2.3.   Clark, A., *The Theory of Adsorption and Catalysis*, Academic Press, New York (1970).
2.4.   de Boer, J. H., *The Dynamic Character of Adsorption*, Oxford University Press, London (1950).
2.5.   Gatos, H., *The Surface Chemistry of Metals and Semiconductors*, John Wiley, New York (1959).
2.6.   Garner, W. E. (Editor), *Chemisorption*, Butterworths Scientific Publications, London (1957).
2.7.   Gregg, S. J., and Sing, K. S. W., *Adsorption, Surface Area and Porosity*, Academic Press, New York (1967).
2.8.   Hepple, P., *Chemisorption and Catalysis*, Elsevier, New York (1970).
2.9.   King, D. A., and Woodruff, D. P. (Editors), *The Chemical Physics of Solid Surfaces and Heterogeneous Catalysis, Volume 1: Clean Solid Surfaces*, Elsevier, New York (1981).
2.10. King, D. A., and Woodruff, D. P. (Editors), *The Chemical Physics of Solid Surfaces and Heterogeneous Catalysis, Volume 2: Adsorption at Solid Surfaces*, Elsevier, New York (1983).
2.11. King, D. A., and Woodruff, D. P. (Editors), *The Chemical Physics of Solid Surfaces and Heterogeneous Catalysis, Volume 3A and 3B: Chemisorption Systems*, Elsevier, New York (1984).
2.12. Linsen, B. G. (Editor), *Physical and Chemical Aspects of Adsorbents and Catalysts*, Academic Press, New York (1970).
2.13. Lowell, S., *Introduction to Powder Surface Area*, John Wiley, New York (1979).
2.14. Ponec, V., Knor, Z., and Cerney, S., *Adsorption on Solids*, Cleveland, CRC Press (1974).
2.15. Ross, S., and Olivier, J. P., *On Physical Adsorption*, John Wiley, New York (1964).
2.16. Ruthven, D. M., *Principles of Adsorption and Adsorption Processes*, John Wiley, New York (1984).
2.17. Selwood, P. W., *Adsorption and Collective Magnetism*, Academic Press, New York (1962).
2.18. Smith, J. R. (Editor), *Theory of Chemisorption*, Springer-Verlag, Berlin (1980).
2.19. Somorjai, G. A., *Principles of Surface Chemistry*, Prentice-Hall, Englewood Cliffs, New Jersey (1972).
2.20. Somorjai, G. A., *Chemistry in Two Dimensions: Surfaces*, Cornell University Press, Ithaca (1981).
2.21. Tomkins, F. C., *Chemisorption of Gases on Metals*, Academic Press, New York (1978).
2.22. Trapnell, B. M. W., *Chemisorption*, Butterworth Scientific Publications, London (1964).
2.23. Trapnell, B. M. W., *The Surface Chemistry of Solids*, Reinhold Publishing Co., New York (1961).

3. MECHANISMS, KINETICS, AND RELATED PHENOMENA

3.1.   Bamford, C. H., and Tipper, C. F. H., *Complex Catalytic Processes*, Elsevier, New York (1978).
3.2.   Boudart, M., *Kinetics of Chemical Processes*, Prentice Hall, Englewood Cliffs, New Jersey (1968).

3.3. Boudart, M., and Djega-Mariadasson, G., *Kinetics of Heterogeneous Reactions*, Princeton University Press, Princeton (1984).
3.4. Tamaru, K., *Dynamic Heterogeneous Catalysis*, Academic Press, New York (1978).

4. DIFFUSION

4.1. Jackson, R., *Transport in Porous Catalysts*, Elsevier, New York (1970).
4.2. Mason, E. A., and Malinauskas, A. P., *Gas Transport in Porous Media, The Dusty-Gas Model*, Elsevier, New York (1983).
4.3. Satterfield, C. N., *Mass Transfer in Heterogeneous Catalysis*, MIT Press, Cambridge, Massachusetts (1970).

5. DESIGN OF CATALYSTS

5.1. Balandin, A. A., *Scientific Selection of Catalysts*, Tel Aviv, Israel, Program for Scientific Translations (1968).
5.2. Trimm, D. L., *Design of Industrial Catalysts*, Elsevier, New York (1980).

6. CATALYST PREPARATION AND MANUFACTURE

6.1. Dallavalle, J. M., *Micromeretics, The Technology of Fine Particles*, Pitman Publishing Corp., New York (1948).
6.2. Delmon, B., Jacobs, P. A., and Poncelet, G. (Editors), *Prepration of Catalysts I*, Elsevier, New York (1976).
6.3. Delmon, B., Grange, P., Jacobs, P. and Poncelet, G. (Editors), *Preparation of Catalysts II*, Elsevier, New York (1979).
6.4. Hench, L. L., and Ulrich, D. R., *Science of Ceramic Chemical Processing*, John Wiley, New York (1986).
6.5. Iler, R., *The Chemistry of Silica*, John Wiley, New York (1979).
6.6. Misra, C. (Editor), *Industrial Alumina Chemicals*, ACS Monograph Series No. 184, American Chemical Society, Washington, D.C. (1986).
6.7. Poncelet, P., Grange, P., and Jacobs, P. A. (Editors), *Preparation of Catalysts III*, Elsevier, New York (1983).
6.8. Poncelet, P., Grange, P., and Jacobs, P. A. (Editors), *Preparation of Catalysts, IV*, Elsevier, New York (1987).
6.9. Stiles, A. V., *Catalyst Manufacture*, Marcel Dekker, New York (1983).
6.10. Sittig, M., *Catalyst Manufacture, Recovery and Use*, Noyes Data Corporation, Park Ridge, New Jersey (1972).
6.11. Sittig, M., *Handbook of Catalyst Manufacture*, Noyes Data Corporation, Park Ridge, New Jersey (1978).

7. CATALYST CHARACTERIZATION AND TESTING

7.1. Anderson, J. R., and Pratt, K. C., *Introduction to Characterization and Testing of Catalysts*, Academic Press, New York (1985).

7.2. Anderson, R. B. (Editor), *Experimental Methods in Catalytic Research*, Vol. 1, Academic Press, New York (1968).

7.3. Anderson, R. B., and Dawson, P. T. (Editors), *Experimental Methods in Catalytic Research*, Vol. 2, Academic Press, New York (1976).

7.4. Anderson, R. B., and Dawson, P. T. (Editors), *Experimental Methods in Catalytic Research*, Vol. 3, Academic Press, New York (1976).

7.5. ASTM Committee D-32, *1985 Annual Book of ASTM Standards*, Vol. 5.03, American Society for the Testing of Materials, Philadelphia (1985).

7.6. Barth, H. G. (Editor), *Modern Methods of Particle Size Analysis*, John Wiley, New York (1984).

7.7. Casper, L. A., and Powell, C. J., *Industrial Applications of Surface Analysis*, ACS Symposium Series No. 199, American Chemical Society, Washington, D.C. (1983).

7.8. Delannay, F. (Editor), *Characterization of Heterogeneous Catalysts*, Marcel Dekker, New York (1984).

7.9. Delgass, W. N., Haller, G. L., Kellerman, R., and Lunsford, J. H., *Spectroscopy in Heterogeneous Catalysis*, Academic Press, New York (1979).

7.10. Delviney, M. L., and Gland, J. L. (Editors), *Catalyst Chatacterization Sciences*, ACS Symposium Series No. 288, American Chemical Society, Washington, D.C. (1985).

7.11. Hair, M. L., *Infrared Spectroscopy in Surface Chemistry*, Edward Arnold Ltd., London (1967).

7.12. Little, L. H., *Infrared Spectra of Adsorbed Species*, Academic Press, New York (1966).

7.13. Parfitt, G. D., and Sing, K. S. W. (Editors), *Characterization of Powder Surfaces*, Academic Press, New York (1976).

7.14. Marcus, P. M., and Jona, F., *Determination of Surface Structure by LEED*, Plenum Press, New York (1984).

7.15. Thomas, J. M., and Lambert, R. M. (Editors), *Characterization of Catalysts*, John Wiley, New York (1980).

8. CATALYST DEACTIVATION

8.1. Delmon, B., and Froment, G. F. (Editors), *Catalyst Deactivation*, Elsevier, New York (1980).

8.2. Hegedus, L. L., and McCabe, R. W., *Catalyst Poisoning*, Marcel Dekker, New York (1984).

8.3. Hughes, R., *Deactivation of Catalysts*, Academic Press, New York (1984).

8.4. Kuczynski, G. C. (Editor), *Sintering and Catalysis*, Plenum Press, New York (1975).

8.5. Kuczynski, G. C. (Editor), *Sintering Processes*, Plenum Press, New York (1980).

8.6. Kuczynski, G. C., Miller, A. E., and Sargent, G. A. (Editors), *Sintering and Heterogeneous Catalysis*, Plenum Press, New York (1983).

8.7. Oudar, J., and Wise, H., *Deactivation and Poisoning of Catalysts*, Marcel Dekker, New York (1985).

9. INDUSTRIAL PROCESSES

9.1. Anderson, R. B., *The Fischer-Tropsch Synthesis*, Academic Press, New York (1984).

9.2. Andrew, S. P. S. (Editor), *Catalyst Handbook*, Springer-Verlag, Berlin (1970).

9.3. Augustine, R. L., *Catalytic Hydrogenation*, Marcel Dekker, New York (1965).

9.4. Boor, J. R., *Zeigler-Natta Catalysts and Polymerizations*, Academic Press, New York (1979).
9.5. Brykowski, F. J. (Editor), *Ammonia and Synthesis Gas*, Noyes Data Corp., Park Ridge, New Jersey (1981).
9.6. Chang, C. D., *Hydrocarbons from Methanol*, Marcel Decker, New York (1983).
9.7. Crynes, B. L. (Editor), *Chemical Reactions as a Means of Separation*, Marcel Dekker, New York (1977).
9.8. Cusumano, J. A., Dalla Betta, R. A., and Levy, R. B., *Catalysis in Coal Conversion*, Academic Press, New York (1978).
9.9. Ford, P. C. (Editor), *Catalytic Acitivation of Carbon Monoxide*, ACS Symposium Series No. 152, American Chemical Society, Washington, D.C. (1981).
9.10. Freifelder, M., *Practical Catalytic Hydrogenation, Techniques and Applications*, John Wiley, New York (1971).
9.11. Gillies, M. T. (Editor), *C1-Based Chemicals from Hydrogen and Carbon Monoxide*, Noyes Data Corp., Park Ridge, New Jersey (1982).
9.12. Herman, R. G. (Editor), *Catalytic Conversions of Synthesis Gas and Alcohols to Chemicals*, Plenum Press, New York (1984).
9.13. Hucknall, D. J., *Selective Oxidation of Hydrocarbons*, Academic Press, New York (1974).
9.14. Jennings, M. S. (Editor), *Catalytic Incineration for Control of Volatile Organic Compound Emissions*, Noyes Data Corp., Park Ridge, New Jersey (1985).
9.15. Leach, B. E. (Editor), *Applied Industrial Catalysis*, Vol. 1, Academic Press, New York (1983).
9.16. Leach, B. E. (Editor), *Applied Industrial Catalysis*, Vol. 2, Academic Press, New York (1983).
9.17. Leach, B. E. (Editor), *Applied Industrial Catalysis*, Vol. 3, Academic Press, New York (1984).
9.18. Kieboom, A. P. G., and van Rantwijk, F., *Hydrogenation and Hydrogenolysis in Synthetic Organic Chemistry*, Delft Univ. Press, Delft (1977).
9.19. Marek, L. F., and Halin, D. A., *The Catalytic Oxidation of Organic Compounds in the Vapor Phase*, The Chemical Catalog Company, New York (1932).
9.20. Nielsen, A., *An Investigation on Promoted Iron Catalysts for the Synthesis of Ammonia*, Haldor Topoe, Vedbaek (1968).
9.21. Pearce, R., and Patterson, W. R. (Editors), *Catalysis and Chemical Processes*, John Wiley, New York (1981).
9.22. Peterson, R. J., *Hydrogenation Catalysts*, Noyes Data Corp., Park Ridge, New Jersey (1977).
9.23. Ranney, M. W., *Desulfurization of Petroleum*, Noyes Data Corp., Park Ridge, New Jersey (1975).
9.24. Rasser, J. C., *Platinum-Iridium Reforming Catalysts*, Delft University Press, Delft (1977).
9.25. Rostrup-Nielsen, J. R., *Steam Reforming Catalysts*, Danish Technical Press Inc., Copenhagen (1975).
9.26. Rylander, R. N., *Catalytic Hydrogenation over Platinum Metals*, Academic Press, New York (1967).
9.27. Satriana, M. J. (Editor), *Hydroprocessing Catalysts for Heavy Oil and Coal*, Noyes Data Corp., Park Ridge, New Jersey (1982).
9.28. Scott, J. (Editor), *Zeolite Technology and Applications*, Noyes Data Corp., Park Ridge, New Jersey (1980).
9.29. Sittig, M., *Catalysts and Catalytic Processes*, Noyes Data Corp., Park Ridge, New Jersey (1967).
9.30. Sittig, M., *Automotive Pollution Control Catalysts and Devices*, Noyes Data Corp., Park Ridge, New Jersey (1977).

9.31. Sokol'skii, D. V., *Hydrogenation in Solutions*, Daniel Davey & Co., London (1964).
9.32. Speight, J. G., *The Desulfurization of Heavy Oils and Residua*, Marcel Dekker, New York (1981).
9.33. Storch, H. H., Golumbic, N., and Anderson, *The Fischer-Tropsch and Related Synthesis*, John Wiley, New York (1951).
9.34. Thomas, C. L., *Catalytic Processes and Proven Catalysts*, Academic Press, New York (1970).
9.35. Twigg, M. V. (Editor), *Catalyst Handbook*, Wolfe Publishing Co., London (1989).
9.36. Vancini, C. A., *Synthesis of Ammonia*, CRC Press, Cleveland (1971).
9.37. Venuto, P. B., and Habib, E. T. Jr., *Fluid Catalystic Cracking with Zeolite Catalysts*, Marcell Decker, New York (1979).
9.38. Wojciechowskii, B. W., *Catalytic Cracking: Catalysts, Chemistry and Kinetics*, Marcel Decker, New York (1987).

10. REACTOR ENGINEERING AND MODELING

10.1. Aris, R., *Elementary Chemical Reactor Analysis*, Prentice-Hall, Englewood Cliffs, New Jersey (1969).
10.2. Bisio, A., and Kabel, R. L., *Scaleup of Chemical Processes*, John Wiley, New York (1985).
10.3. Butt, J. B., *Reaction Kinetics and Reactor Design*, Prentice-Hall, Englewood Cliffs, New Jersey (1980).
10.4. Carberry, J. J., *Chemical and Catalytic Reaction Engineering*, McGraw-Hill, New York (1976).
10.5. Cooper, A. R., and Jeffreys, G. V., *Chemical Kinetics and Reactor Design*, Prentice-Hall, Englewood Cliffs, New Jersey (1971).
10.6. Denbigh, K. G., *Chemical Reactor Theory*, Cambridge Press, Cambridge (1966).
10.7. Doraiswamy, L. K., and Mashelkar, R. A., *Frontiers in Chemical Reaction Engineering*, Vols. 1 and 2, John Wiley, New York (1984).
10.8. Dudukovic, M. P., and Mills, P. L., *Chemical and Catalytic Reactor Modeling*, ACS Symposium Series No. 237, American Chemical Society, Washington, D.C. (1983).
10.9. Fogler, H. S., *Elements of Chemical Reaction Engineering*, Prentice-Hall, Englewood Cliffs, New Jersey (1986).
10.10. Froment, G. F., and Bischoff, K., *Chemical Reactor Analysis and Design*, John Wiley, New York (1979).
10.11. Gianetto, A., and Silveston, P. L. (Editors), *Multiphase Chemical Reactors, Theory, Design, Scale-up*, Hemisphere Publishing Corp., New York (1986).
10.12. Hill, C. G., *Introduction to Chemical Engineering Kinetics and Reactor Design*, John Wiley, New York (1977).
10.13. Horak, J., and Pasek, J., *Design of Industrial Chemical Reactors from Laboratory Data*, Heyden, Prague (1978).
10.14. Lapidus, L., and Amundson, N. R. (Editors), *Chemical Reactor Theory, A Review*, Prentice-Hall, Englewood Cliffs, New Jersey (1977).
10.15. Lee, H. L., *Heterogeneous Reactor Design*, Butterworths, London (1984).
10.16. Levenspiel, O., *Chemical Reaction Engineering*, John Wiley, New York (1972).
10.17. Matros, Yu., Sh. (Editor), *Unsteady Processes in Catalytic Reactors, Studies in Surface Science and Catalysis, 22,* Elsevier, New York (1985).
10.18. Peterson, E. E., *Chemical Reaction Analysis*, Prentice-Hall, Englewood Cliffs, New Jersey (1965).

10.19. Prins, R., and Schuit, G. C. A. (Editors), *Chemistry and Chemical Engineering of Catalytic Processes*, Sijthoff and Noordhoff, Germantown (1980).
10.20. Rase, H. F., *Chemical Reactor Design for Process Plants*, Vols. 1 and 2, John Wiley, New York (1977).
10.21. Rose, L. M., *Chemical Reactor Design in Practice*, Elsevier, New York (1981).
10.22. Shah, Y. T., *Reaction Engineering in Direct Coal Liquefactions*, Addison-Wesley Publishing Co., Reading, Massachusetts (1981).
10.23. Smith, J. M., *Chemical Engineering Kinetics*, McGraw-Hill, New York (1980).
10.24. Tarhan, M. O., *Catalytic Reactor Design*, McGraw-Hill, New York (1983).
10.25. Westerterp, K. R., van Swaaij, W. P. M., and Beenackers, A. A. C. M., *Chemical Reactor Design and Operation*, John Wiley, New York (1984).

11. PROCESS DESIGN

11.1. Aerstin, F., and Street, G., G., *Applied Chemical Process Design*, Plenum Press, New York (1978).
11.2. Benn, F. R., Edewar, J. O., and McAuliffe, C. A., *Production and Utilization of Synthetic Fuels*, John Wiley, New York (1981).
11.3. Cooler, B. R. (Editor), *The Science and Technology of Coal and Coal Utilization*, Plenum Press, New York (1984).
11.4. Gary, J. H., and Handwerk, G. L., *Petroleum Refining, Technology and Economics*, Marcel Dekker, New York (1975).
11.5. Hahn, A. V. G., *The Petroleum Industry: Market and Economics*, McGraw-Hill, New York (1970).
11.6. Hobson, G. D., and Pohl, W (Editors), *Modern Petroleum Technology*, Part 1 and 2, John Wiley, New York (1984).
11.7. Holland, F. A., Watson, F. A., and Wilkinson, J. K., *Introduction to Process Economics*, John Wiley, New York (1984).
11.8. Husain, A., *Chemical Process Simulation*, John Wiley, New York (1986).
11.9. Jawad, M. H., and Farr, J. R., *Structural Analysis and Design of Process Equipment*, John Wiley, New York (1984).
11.10. Jelel, F. C. (Editor), *Cost and Optimization Engineering*, McGraw-Hill, New York (1970).
11.11. Kenney, W. F., *Energy Conservation in the Process Industries*, Academic Press, New York (1984).
11.12. Newman, S. A., *Shale Oil Upgrading and Refining*, Butterworths, London (1983).
11.13. Mangold, E., Muradaz, M. A., and Cheremisinoff, P. N., *Coal Liquefaction and Gasification Technologies*, Butterworths, London (1982).
11.14. Mecklenburgh, J. C., *Process Plant Layout*, John Wiley, New York (1985).
11.15. Rider, D. K., *Energy: Hydrocarbon Fuels and Chemical Resources*, John Wiley, New York (1981).
11.16. Sharp, D. H., and West, T. F. (Editors), *The Chemical Industry*, John Wiley, New York (1982).
11.17. Shreve, R. N., and Brink, Jr., J. A., *Chemical Process Industries*, McGraw-Hill, New York (1977).
11.18. Speight, J. G., *The Chemistry and Technology of Petroleum*, Marcel Dekker, New York (1980).
11.19. Stratton, A. (Editor), *Energy and Feedstocks in the Chemical Industries*, John Wiley, New York (1983).
11.20. Waddams, A. L., *Chemicals from Petroleum*, John Wiley, New York (1973).

11.21. Wei, J., Russell, T. W. F., and Swartzlander, M. W., *The Structure of the Chemical Processing Industries*, McGraw-Hill, New York (1979).
11.22. Wiseman, P., *Petrochemicals*, John Wiley, New York (1986).

12. CONTINUING SERIES

12.1. *Catalysis*, edited by P. H. Emmett, Reinhold Publishing Co., New York, Vol. 1 (1954) through vol. 7 (1960).
12.2. *Advances in Catalysis*, edited by d. D. Eley, H. Pines, and P. B. Weisz, Academic Press, New York, Vol. 1 (1948) through Vol. 35 (1987).
12.3. *Catalysis Reviews, Science and Engineering*, edited by A. T. Bell and J. J. Carberry, Marcel Dekker, New York, Vol. 1 (1968) through Vol. 30 (1988).
12.4. *Catalysis*, edited by D. A. Dowden, C. Kemball, G. C. Bond, and G. Webb, The Royal Society of Chemistry, London, Vol. 1 (1977) through Vol. 7 (1985).
12.5. *Catalysis, Science and Technology*, edited by J. R. Anderson and M. Boudart, Springer-Verlag, Berlin, Vol. 1 (1981) through Vol. 7 (1985).
12.6. *Advances in Chemical Engineering*, edited by J. Wei, Academic Press, New York, Vol. 1 (1956) through Vol. 14 (1988).
12.7. *Surface and Colloid Science*, edited by E. Matijevic, Plenum Press, New York, Vol. 1 (1968) through Vol. 14 (1987).

13. PROCEEDINGS (NOT LISTED PREVIOUSLY)

13.1. *Proceedings of the Second International Congress on Catalysis, Paris, 1960*, Vols. I and II, Technip, Paris (1961).
13.2. *Proceedings of the Third International Congress on Catalysis, Amsterdam, 1964*, Vols. I and II, edited by W. M. H. Sachtler, G. C. A. Schuit, and P. Zweitering, John Wiley, New York (1965).
13.3. *Proceedings of the Fourth International Congress on Catalysis, Moscow, 1968*, edited by J. W. Hightower, Rice University, Houston (1969).
13.4. *Proceedings of the Fifth International Congress on Catalysis, Miami Beach, 1972*, edited by J. W. Hightower, Elsevier, New York (1973).
13.5. *Proceedings of the Sixth International Congress on Catalysis, London, 1976*, The Chemical Society, London (1977).
13.6. *Proceedings of the Seventh International Congress on Catalysis, Tokyo, 1980*, edited by T. Seiyama and K. Tanabe, Elsevier, Amsterdam (1981).
13.7. *Proceedings of the Eighth International Congress on Catalysis, Berlin, 1984*, not published.
13.8. *Proceedings of the 25th International Meeting of the Societe de Chimie Physique, Reaction Kinetics in Chemical Systems, Dijon, 1974*, edited by P. Parret, Elsevier, New York (1974).
13.9. *Catalysis, Heterogenous and Homogeneous, Proceedings of the International Symposium on the Relations between Heterogeneous and Homogeneous Catalytic Phenomena, Brussels, 1974*, edited by B. Delmon and G. Jannes, Elsevier, Amsterdam (1975).
13.10. *Growth and Properties of Metal Clusters, Proceedings of the 32nd International Meeting of the Societe de Chimie Physique, Villeurbanne, 1979*, edited by J. Bourdon, Elsevier, Amsterdam (1980).
13.11. *Catalysis by Zeolites, Proceedings of an International Symposium, Lyon, 1980*, edited by B. Imelik, C. Naccache, Y. Ben Taarit, J. C. Vedrine, G. Coudurier, and H. Praliaud, Elsevier, Amsterdam (1980).

13.12. *Physics of Solids, Proceedings of the Symposium, Bechyne, Czechoslovakia, 1980,* edited by M. Laznicka, Elsevier, Amsterdam (1982).

13.13. *Adsorption at the Gas–Solid and Liquid–Solid Interface, Proceedings of an International Symposium, Aix-en-Provence, 1981,* edited by J. Rouquerol and K. S. W. King, Elsevier, Amsterdam (1982).

13.14. *Metal-Support and Metal-Additive Effects in Catalysis, Proceedings of an International Symposium, Lyon, 1982,* edited by B. Imelik, C. Naccache, G. Coudurier, H. Praliaud, P. Meriaudeau, P. Gallezot, G. A. Martin, and J. C. Vedrine, Elsevier, Amsterdam (1982).

13.15. *Metal Microstructures in Zeolites, Proceedings of a Workshop,* edited by P. A. Jacobs, N. I. Jaeger, P. Jiru and G. Schulz-Ekloff, Elsevier, Amsterdam (1982).

13.16. *Vibrations at Surfaces, Proceedings of the Third International Conference, Asilomar, California, 1982,* edited by C. R. Brundle and H. Morawitz, Elsevier, Amsterdam (1983).

13.17. *Proceedings of the Robert A. Welch Foundation Conferences on Chemical Research XXV. Heterogeneous Catalysis, 1981,* edited by W. O. Milligan, The Robert A. Welch Foundation, Houston, Texas (1983).

13.18. *Spillover of Adsorbed Species, Proceedings of the International Symposium, Lyon-Villeur-banne, 1983,* edited by G. M. Pajonk, S. J. Teichner, and J. E. Germain, Elsevier, Amsterdam (1983).

13.19. *Structure and Reactivity of Modified Zeolites, Proceedings of an International Conference, Prague, 1984,* edited by P. A. Joacobs, N. I. Jaeger, P. Jiru, V. B. Kazansky, and G. Schulz-Ekloff, Elsevier, Amsterdam (1984).

13.20. *Catalysis on the Energy Scene, Proceedings of the 9th Canadian Symposium on Catalysis, Quebec, 1984,* edited by S. Kaliaguine and A. Mahay, Elsevier (1984).

14. JOURNALS

14.1. *Journal of Catalysis,* edited by W. K. Hall and F. S. Stone, Vol. 1 (1962)—, Academic Press, New York.

14.2. *Applied Catalysis,* edited by B. Delmon, regional editors: J. Roth, L. Guczi, D. L. Trimm, J. C. Vedrine, and D. A. Whan, Vol. 1 (1981)—, Elsevier, New York.

14.3. *Journal of Molecular Catalysis,* edited by E. G. Derouane, W. Marconi, C. U. Pittman, Jr., Ph. Teyssie, and W. R. Vieth, Vol. 1 (1976)—, Elsevier, New York.

14.4. *Kinetics and Catalysis,* edited by T. J. Gray, Vol. 1 (1960)—, Consultants Bureau, New York.

14.5. *Reaction Kinetics and Catalysis Letters,* edited by G. K. Boreskov and F. Nagy, Vol. (1985)—, Elsevier, Amsterdam.

14.6. *Catalyst Reports,* edited by J. R. H. Ross, Vol. 1 (1987)—, Elsevier, Amsterdam.

14.7. *Journal of Colloid and Interface Science,* edited by M. Kerker and A. C. Zettlemoyer, Vol. 1 (1971)—, Academic Press, York.

14.8. *Colloids and Surfaces,* edited by P. Somasundaran, Vol. 1 (1979)—, Elsevier, New York.

14.9. *Surface Science,* edited by H. C. Gatos, Vol. 1 (1964)—, North-Holland Publishing Co., Amsterdam.

14.10. *Progress in Surface Science,* edited by S. G. Davison, Vol. 1 (1971)—, Pergamon Press, New York.

14.11. *Reactivity of Solids,* edited by P. Barret, P. K. Gallagher, and J. Haber, Vol. 1 (1985)—, Elsevier, New York.

14.12. *Industrial and Engineering Chemistry, Product Research and Development,* edited by J. A. Seiner, Vol. 17 (1978)—, American Chemical Society, Washington, D.C.

14.13. *Industrial and Engineering Chemistry, Process Design and Development,* edited by H. M. Gulburt, Vol. 1 (1962)—, American Chemical Society, Washington, D.C.

14.14. *Industrial and Engineering Chemistry, Fundamental,* edited by R. L. Pigford, Vol. 1 (1962)—, American Chemical Society, Washington, D.C.

14.15. *Energy and Fuels,* edited by J. W. Larsen, Vol. 1 (1987)—, American Chemical Society, Washington, D.C.

14.16. *AICHE Journal,* edited by M. M. Denn, Vol. 1 (1955)—, American Institute of Chemical Engineers, New York.

14.17. *The Canadian Journal of Chemical Engineering,* edited by N. Epstein, Vol. 1 (1957)—, Canadian Society of Chemical Engineering, Toronto.

14.18. *Chemical Week,* edited by P. P. McCurdy, Vol. 1 (1951)—, McGraw-Hill, New York.

14.19. *Chemical Engineering Progress,* edited by A. K. Dubberly, Vol. 1 (1908)—, American Institute of Chemical Engineers, New York.

APPENDIX 2

Glossary of Common Terms in Refining and Petrochemicals

Activity   A measure of the effectiveness of a catalyst, usually (1) rate at standard conditions, (2) rate constant, (3) activation energy or preexponential factor, (4) conversion, (5) space velocity for a given conversion, or (6) temperature for a given conversion.

Alkylates   Product of alkylation processes, usually high-octane, branched-chain hydrocarbons.

Alkylation   Upgrading of hydrocarbons by replacing hydrogen with an alkyl radical.

Aluminum Chloride Treating   Improvement of steam-cracked naphthas using aluminum chloride as a catalyst; improves color and odor.

Analine Point   A test of the paraffinicity of a sample in which the minimum temperature for complete miscibility of analine and the sample is measured.

Aromatic Cycloparaffins   Ring hydrocarbons containing a benzene ring attached to a cycloparaffin through two adjacent carbon atoms.

Aromatization   Formation of an aromatic usually from an alicyclic hydrocarbon.

ASTM Distillation   A standardized laboratory bath distillation test for naphthas and middle distillates at atmospheric pressure without fractionation.

Barrel   A measure of petroleum equal to 42 U.S. gallons.

Barrels per Calendar Day (BPCD)   Average flow rate based on 365 days per year.

Barrels per Stream Day (BPSD)   Average flow rate based on actual on-stream time of a unit.

Battery Limits   The periphery of a process unit, including all equipment for the process.

Bicycloparaffins   Two-ring, bonded cycloparaffins with five- or six-membered rings.

Bitumen   The part of a petroleum, asphalt, and tar product that dissolves in carbon disulfide.

Blending   Mixing of two or more components to achieve desired properties of the finished product.

Bromine Index   Measure of the amount of bromine-reactive material in a sample.

Bromine Number   A test for the degree of unsaturation in a sample.

Bottoms   The higher-boiling residue removed from a distillation column.

Caffeine Number   A measure of the amount of carcinogenic compounds in an oil.

Carbanion Ion   A negatively charged hydrocarbon ion, e.g., RCH^-.

Carbonion Ion   A positively charged hydrocarbon ion, e.g., RCH^+.

Catalytic Cracking   A process in which higher-molecular-weight hydrocarbons are converted to lighter compounds, thereby producing an improvement in quality.

Catalytic Reforming   Conversion of naphtha fractions to products of higher octane value; principally isomerization, cyclization dehydrogenation, and hydrocracking.

Cetene Number   A measure of the ignition quality of diesel fuel.

Cloud Point   The temperature at which solidifiable compounds crystallize in a sample.

Complete Combustion   Complete oxidation of a hydrocarbon giving CO_2 and H_2O as the products.

Condensation   Combination of two different organic molecules, with or without elimination of H_2O or HCl.

Conradson Carbon   A test of the amount of carbon residue left after evaporization and pyrolysis of an oil.

Copolymer   Polymer produced from two or more different monomers.

Cracking   Rupture of C–C bonds in hydrocarbons to give smaller, less saturated fragments without loss or gain of hydrogen.

Cut   That portion of a crude boiling within certain temperature limits.

Cyclization   Production of an alicyclic from an aliphatic hydrocarbon.

Cycloparaffins   Saturated ring hydrocarbons, with five or six carbons to each ring.

Dearomatization   Reverse of aromatization.

Decyclization   Formation of an aliphatic from an alicyclic compound.

Dehalogenation   Reverse of halogenation.

Dehydration   The reverse of hydration, involving the splitting off of an active hydrogen to give water or ammonia.

Dehydrogenation   The reverse of hydrogenation, involving the splitting off of hydrogen to give a less saturated compound and the rupture of a C–H bond.

Demetallization   Removal of metal (iron, nickel, and vanadium) porphorins by a combination of hydrotreating and adsorption.

Desulfurization   Same as hydrodesulfurization.

Dinuclear Aromatics   Fusion of two benzene rings with two common carbon atoms.

Dismutation   The same as disproportionation.

Disproportionation   Splitting of a hydrocarbon into two different hydrocarbons, with one a lower and the other a higher carbon number and no gain or loss of hydrogen.

Distillation   A mechanical process for the separation, under atmospheric pressure or vacuum, of crude oil into various boiling point fractions.

Doctor Test   A test for mercaptain sulfur in petroleum products.

Dual Oxygenolysis   Rupture of C–C and C–H bonds to give both H_2O and CO_2.

Fischer–Tropsch Reactions   Reaction of CO and H_2 to produce paraffins, olefins, alcohols, aldehydes, ketones, and fatty acids.

Fixed Carbon   The organic portion of residual coke obtained on evaporation to dryness of hydrocarbon products in the absence of air.

Free Carbon   Organic matter in tars that is insoluable in carbon disulfide.

Formulation   Production of an aldehyde or ketone from a hydrocarbon and carbon monoxide without using hydrogen.

Gas Hourly Space Velocity (GHSV)   Volume of gas (STP) per volume of catalyst per hour.

Gas Oil   Petroleum fractions boiling in the ranges 200–350°C (light), 325–400°C (medium), and 380–575°C (vacuum).

Halogenation   Addition of a halogen atom (F, Cl, Br, and I) to a molecule.

Heat Transfer Limited Reaction   An endothermic reaction whose rate is determined by the rate of heat transfer to the reactor.

Hydration   Addition of H_2O or a compound containing active hydrogen, such as NH_3.

Hydrocarboxylation   Formation of an acid from an alkene, CO, and H_2O.

Hydrocracking   A dual-type reaction of cracking and hydrogenation to produce saturates of high octane value from heavier compounds. It is accompanied by the hydrogenolysis of C–S, C–N, and C–O.

Hydrodealkylation   Removal of an alkyl side chain from an alkyl aromatic by substituting hydrogen.

Hydrodenitrogenation   Removal of nitrogen as NH_3 by the addition of hydrogen.

Hydrodesulfurization   Removal of sulfur as H_2S by the addition of hydrogen.

Hydroformylation   Addition of hydrogen to produce formylation.

Hydrogenation   Addition of hydrogen to unsaturated bonds.

Hydrogenolysis   Used in two different ways: (1) hydrogenation of an oxygen compound to give a reduced compound and water, and (2) rupture of C-C, C-S, C-N, C-O, and other hetero C- bonds by the addition of hydrogen.

Hydrogen Transfer   A reaction involving the rupturing of a C-H bond and the transfer of the hydrogen atom to another molecule.

Hydrotreating   General process of adding hydrogen to remove S, N, and O; improves color, stability, odor, and further processing.

Isomerization   Rearrangement of the structure of a compound without gain or loss of any of its components.

Isotope Exchange   Exchange of isotopes of different mass numbers from specific parts of a molecule, permitting tracing of specific atoms to elucidate mechanisms.

Kerosine   A middle-distillate material boiling in the range 150-260°C.

Lifetime   A measure of the effective life of a catalyst, usually process time or amount of feed per unit of catalyst.

Liquid Hourly Space Velocity (LHSV)   Liquid volume (normal conditions) per volume of catalyst per hour.

Liquefied Petroleum Gas (LPG)   Liquefied light end gases used as fuel and chemical feedstock, usually 95% propane, the rest ethane and butane.

Mass Transfer Limited Reaction   A reaction whose rate is determined by a mass transfer process, such as external diffusion to the outside of a pellet or pore diffusion into the interior surface.

Metathesis   General term for reactions in which parts of two compounds exchange with each other.

Methanation   Hydrogenation of carbon monoxide and carbon dioxide to give methane and water.

Mid-Boiling Point   The temperature in a crude assay distillation at which one half of the material of a cut has vaporized.

Middle Distillate   Atmospheric pipestill cuts boiling in the range 150-400°C.

Monocycloparaffins   Cycloparaffins containing only one ring, of normally five or six carbon atoms.

Monomer   A simple compound containing unsaturated bonds and capable of joining to itself or other molecules to form larger polymers.

Motor Octane Number (MON)   The percentage by volume of isooctane in a blend of isooctane and *n*-heptane that knocks with the same intensity as fuel being tested at conditions approximating cruise conditions of an automobile (900 rpm).

Naphtha   Petroleum fractions in the boiling ranges 25-75°C (light virgin), 75-140°C (intermediate), 140-165°C (heavy), and 165-220°C (very heavy).

Naphthenes   Same as cycloparaffins.

Natural Gas Liquids (NGL)   Liquefied hydrocarbons, usually ethane and propane, separated from natural gas.

Oxidation   Reaction involving the breaking of a C–C or C–H bond with oxygen.

Oxychlorination   Chlorination with addition of oxygen to prevent loss of chlorine as HCl.

Oxydehydrogenation   Removal of hydrogen from a compound by reaction with oxygen to produce water.

Oxygenolysis   Rupture of C–H, C–C, C–S, C–N by oxygen to produce H_2O, CO_2, SO_2, and NO_2.

Partial oxidation   Oxidation to produce an oxidized organic compound rather than H_2O and CO_2.

Polymer   Larger molecule formed from monomers.

Polymerization   Process for making polymers.

Polynuclear Aromatics   Compounds containing two or more benzene rings are fused through common carbon atoms.

Pour Point   The lowest temperature at which a petroleum product will flow.

Raffinate   The residue recovered from an extraction process.

Reductive Alkylation   Alkylation of an amine with a ketone in the presence of hydrogen to give a secondary amine and water.

Refining   A general term that includes processing of crude oils to give fuels such as gasoline, diesel and jet, lubricants, asphalts, waxes, and chemical feedstocks.

Reformate   A reformed naphtha upgraded in octane number by catalytic reforming.

Research Octane Number (RON)   The percentage by volume of isooctane in a blend of isooctane and *n*-heptane that knocks with the same intensity as a gasoline being tested under conditions approximating an automobile at low speeds (600 rpm).

Shift Reaction   Reaction of steam with carbon-containing materials to give carbon dioxide and hydrogen.

Selectivity   Used in two ways: (1) the ratio of the rates of formation of a desired and undesired product, and (2) the ratio of the amount of desired and undesired product.

Sensitivity   The difference between the research and the motor octane numbers.

Side Chain   An aliphatic group attached to a hydrocarbon.

Smoke Point   A test of the burning quality of jet fuels, kerosene and alliminating oils.

Sour or Sweet Crude   A general term for classifying crudes according to sulfur content: sour crude—greater than 0.5 to 1%; sweet crudes—less than 0.5%.

Steam Cracking   A noncatalytic process in which hydrocarbons are cracked in the presence of steam at elevated temperatures.

Steam Reforming   Reaction of steam with carbon-containing materials to give carbon monoxide and hydrogen.

Straight Run Gasoline   An uncracked gasoline fraction distilled from crude, containing most paraffinic hydrocarbons.

Substitute Natural Gas (SNG)   Methane produced from synthesis gas derived from steam reforming naphthas, coal, or biomass.

Synthesis Gas   Any mixture of hydrogen and carbon monoxide.

Transalkylation   An alkylation reaction with an intramolecular alkyl group transfer.

Visbreaking   Thermal cracking of crude oil or reduced crude under mild conditions.

Weight Hourly Space Velocity   Mass of feed per mass of catalyst per hour.

Yield   Use in two ways: (1) the ratio of the amount of a given product to the amount of reactant; and (2) the ratio of the amount of a given product to the amount of reactant converted.

APPENDIX 3
Units and Nomenclature

Quantity	Symbol	Common units	SI Units	Conversion factor
1. DIFFUSION				
Bulk diffusivity	D_B	cm^2/s	$m^2\,s^{-1}$	$1.000E-4$
Knudsen diffusivity	D_K	cm^2/s	$m^2\,s^{-1}$	$1.000E-4$
Pellet effective diffusivity	D_{eff}	cm^2/s	$m^2\,s^{-1}$	$1.000E-4$
Heat transfer coefficient	h	$cal/s\,cm^2\,°C$	$J\,s^{-1}\,m^{-2}\,K^{-1}$	$4.187E+4$
Heat transfer factor	j_H	—	—	—
Mass transfer coefficient	k_c	cm/s	ms^{-1}	$1.000E-2$
	k_g	$moles/s\,cm^2\,atm$	$mole\,m^{-2}\,s^{-1}\,Pa^{-1}$	$9.871E-8$
Mass transfer factor	j_D	—	—	—
Thiele modulus	Φ	—	—	—
Effective factor	η, EF	—	—	—
2. ADSORPTION				
Volume adsorbed	V	cm^3/g	$m^3\,kg^{-1}$	$1.000E-3$
Monolayer volume	V_m	cm^3/g	$m^3\,kg^{-1}$	$1.000E-3$
Surface area	S_v	m^2/cm^3 (cat)	$m^2(m^3\,cat)^{-3}$	$1.000E+6$
	S_g	m^2/g	$m^2\,kg^{-1}$	$1.000E+3$
Surface coverage	θ	—	—	—
Adsorption coefficient	K_A	atm^{-1}	Pa^{-1}	$9.869E-6$
Enthalpy of adsorption	H_a	$kcal/mole$	$J\,mole^{-1}$	$4.187E+3$
Activation energy of adsorption	E_a	$kcal/mole$	$J\,mole^{-1}$	$4.187E+3$
3. KINETICS				
Total pressure	P_T	atm	Pa	$1.013E+5$
Partial pressure	P_A	atm	Pa	$1.013E+5$
Concentration	C_A	mole/liter	$mole\,m^{-3}$	$1.000E+3$

(continued)

Quantity	Symbol	Common units	SI Units	Conversion factor
Mass of catalyst	W	g	kg	$1.000E+3$
Volumetric flow rate	F_0	cm^3/s	$m^3 s^{-1}$	$1.000E-6$
Contact time	τ	s	s	$1.000E00$
Space velocity	S	s^{-1}	s^{-1}	$1.000E00$
Gas hourly space velocity	GHSV	cm^3/cm^3 (cat)/hr	s^{-1}	$2.778E-4$
Liquid hourly space velocity	LHSV	cm^3/cm^3 (cat)/hr	s^{-1}	$2.778E-4$
Fractional conversion	X_A	—	—	—
Reaction rate	$-R_A$	mole/cm^3 (cat) s	mole $m^{-3} s^{-1}$	$1.000E+3$
Turnover number	N	mol/site sec	mol/site$^{-1} s^{-1}$	$1.000E00$
Selectivity	S	—	—	—
Yield	Y	—	—	—
Order of reaction	n, m	—	—	—
First-order rate	k_v	s^{-1}	s^{-1}	$1.000E00$
constant	k_g	cm^3/g s	$m^3 kg^{-1} s^{-1}$	$1.000E-3$
	k_s	cm/s	ms^{-1}	$1.0000E-2$
Activation energy	E	kcal/mole	J mole^{-1}	$4.187E+3$

4. CATALYST PROPERTIES

Quantity	Symbol	Common units	SI Units	Conversion factor
Particle radius	R_p	cm	m	$1.000E-2$
Particle volume	V_c	cm^3	m^3	$1.000E-6$
Pore volume	V_g	cm^3/g	$m^3 kg^{-1}$	$1.000E-3$
Macro porosity	θ_M	—	—	—
Meso porosity	θ_μ	—	—	—
Average pore radius	r_e	A^0	nm	$1.000E-1$
Specific heat	C_p	cal/g °C	J kg^{-1} K^{-1}	$4.187E+3$
Thermal conductivity	k	cal/cm s °C	Jm^{-1} s^{-1} K^{-1}	$4.187E+2$
Theoretical density	d_T	g/cm^3 (solid)	kgm^{-3}	$1.000E+3$
Particle density	d_p	g/cm^3 (particle)	kgm^{-3}	$1.000E+3$
Packing density	d_B	g/cm^3 (bed)	kgm^{-3}	$1.000E+3$
External pellet area	S_{ext}	m^2/g (particle)	m^2 kg^{-1}	$1.000E+3$
Crushing strength	F	kgf/cm^2	Pa	$9.807E+4$

5. CATALYTIC REACTORS

Quantity	Symbol	Common units	SI Units	Conversion factor
Reactor length	L	cm	m	$1.000E-2$
Reactor radius	R	cm	m	$1.000E-2$
Reactor volume	V_R	cm^3	m^3	$1.000E-6$
Fluid velocity	U	cm/s	ms^{-1}	$1.000E-2$
Mass velocity	G	g/s cm^2 (total)	kgs^{-1} m^{-2}	$1.000E+1$
Fluid viscosity	μ	poise	Nsm^{-2}	$1.000E-1$
Fluid thermal conductivity	k	cal/cm s °C	Jm^{-1} s^{-1} K^{-1}	$4.187E+2$
Fluid heat capacity	C_p	cal/mole °C	J mole^{-1} K^{-1}	$4.187E00$
Fluid density		g/cm^3	kgm^{-3}	$1.000E+3$
Void fraction	ε	—	—	—
Reynolds number	N_{Re}	—	—	—
Schmidt number	N_{Sc}	—	—	—
Prandtl number	N_{Pr}	—	—	—

Quantity	Symbol	Common units	SI Units	Conversion factor
Temperature	T	°C	K	°C + 273
6. PETROLEUM REFINING				
Barrel	BBL	42 US gal	m^3	$1.590E-1$
Throughput	MMBD	Millions of barrels per day	$m^3\,s^{-1}$	$1.840E00$

APPENDIX 4
Catalytic Process Assessment

1. Need for the process
    1.1. Objectives
        1.1.1. Technical
        1.1.2. Economic
        1.1.3. Social
    1.2. Markets
        1.2.1. Past
        1.2.2. Future
        1.2.3. Competing processes

2. Historical developments

3. Feedstocks
    3.1. Supply
        3.1.1. Sources
        3.1.2. Volumes
        3.1.3. Transportation
    3.2. Composition
        3.2.1. Main components
        3.2.2. Impurities
    3.3. Alternate feedstocks

4. Products
    4.1. Principal
    4.2. Coproducts
    4.3. Byproducts
    4.4. Potential separation problems

5. Chemistry
   5.1. Main reactions
   5.2. Side reactions
   5.3. Thermodynamics
        5.3.1. Equilibrium conversions
        5.3.2. Reaction enthalpies
   5.4. Mechanisms
        5.4.1. Molecular
        5.4.2. Surface

6. Catalytic functions
   6.1. Desired activities
   6.2. Activities to be inhibited

7. Catalyst components
   7.1. Active components
        7.1.1. Metals
        7.1.2. Oxides or sulfides
        7.1.3. Acid oxides
   7.2. Supports
   7.3. Promoters
        7.3.1. For the active component
        7.3.2. For the support

8. Proven catalysts
   8.1. Types
   8.2. Commercial nomenclature

9. Catalyst characteristics (for each of the listed catalysts)
   9.1. Name
   9.2. Composition
   9.3. Structure or phases
   9.4. Preparation conditions
        9.4.1. Method
        9.4.2. Materials
        9.4.3. Precipitation or other
        9.4.4. Washing
        9.4.5. Drying
        9.4.6. Calcination
   9.5. Formulation
        9.5.1. Method
        9.5.2. Shape
        9.5.3. Size

9.6. Mechanical properties
    9.6.1. Crushing strength (how measured?)
    9.6.2. Loss on attrition
    9.6.3. Loss on ignition
    9.6.4. Shrinkage
9.7. Particle properties
    9.7.1. Theoretical density
    9.7.2. Pellet density
    9.7.4. Porosity
    9.7.5. Bed density
    9.7.6. Void fraction
9.8. Texture properties
    9.8.1. BET Surface area
    9.8.2. Pore volume
    9.8.3. Average pore size
    9.8.4. Pore size distribution
9.9. Active component
    9.9.1. Phase or structure
    9.9.2. Dispersion
9.10. Acidity
9.11. Activation
9.12. Performance properties
    9.12.1. Method
    9.12.2. Activity
    9.12.3. Selectivity
    9.12.4. Lifetime
    9.12.5. Regeneration
9.13. Potential performance problems
    9.13.1. Mechanical
    9.13.2. Chemical
    9.13.3. Diffusional
    9.13.4. Deactivation
    9.13.5. Regeneration
9.14. Catalyst economics
    9.14.1. Cost
    9.14.2. Availability
    9.14.3. Transportation
9.15. Catalyst loading
    9.15.1. Special handling features
    9.15.2. Safety factors
    9.15.3. Disposal or reclaiming problems
9.16. Legal responsibilities

9.17. Catalytic mechanisms
9.18. Kinetics and rate equations

10. Reactor technology
    10.1. Type of reactor
    10.2. Size and construction
    10.3. Start-up procedures
    10.4. Operating conditions
    10.5. Performance
    10.6. Deactivation
    10.7. Regeneration procedures
    10.8. Shut-down procedures
    10.9. Operating problems

11. Process integration
    11.1. Process design
    11.2. Process configuration
    11.3. Auxiliary units
    11.4. Process economics

12. Future developments
    12.1. Changing markets
    12.2. Changing feedstocks
    12.3. Improved catalysts
    12.4. Improved processes
    12.5. Alternate processes

APPENDIX 5
Classification of Reactions

1. Single-Molecule Reactions
   1.1. Cracking. Rupture of a C—C bond, normally of larger compounds, to give smaller, less saturated compounds.
      1.1.1. Aliphatics

Alkane:	$C_6H_{14} = C_2H_4 + C_4H_{10}$	(dealkylation)
Alkene:	$C_4H_8 = 2C_2H_4$	(depolymerization)

   1.1.2. Alicyclics

   $$C_3H_5 = CH_3CHCH_2 \qquad \text{(dealkylation)}$$

   1.1.3. Aromatics

   $$C_6H_6C_3H_7 = C_3H_6 + C_6H_6 \qquad \text{(dealkylation)}$$

   1.2. Isomerization. The intramolecular rearrangement of structure.
      1.2.1. Aliphatics

   Alkane: $C_4H_{10} = (CH_3)_2CHCH_3$

   Alkene:

   $$CH_3CH_2CH{=}CH_2 = CH_3CH{=}CHCH_3$$

   (shift of double bond)

   $$CH_3CH_2CH{=}CHCH_3 = CH_3CH_2CH{=}CHCH_3$$

   (*cis-trans* shift)

   $$CH_3CH_2H{=}CH_2 = (CH_3)_2C{=}CH_2 \qquad \text{(tertiary C atoms)}$$

   $$(CH_3)_3C{-}CH{=}CH_2 = (CH_3)_2CH{-}C(CH_3){=}CH_2$$

   (tertiary C atom position)

1.2.2. Alicyclics

$$C_5H_9CH_3 = C_6H_{12}$$

1.2.3. Aromatics

$$C_6H_5C_2H_5 = C_6H_4(CH_3)_2 \qquad \text{(mono- to disubstituted)}$$

$$C_6H_5C_4H_9 = C_6H_5CH_2CH(CH_3)_2 \qquad \text{(side-chain rearrangement)}$$

2. Combination Reactions
   2.1. Hydrogen transfer. Rupture of a C–H bond and transfer of the hydrogen atom to another molecule.
      2.1.1. Aliphatics

$$2C_4H_8 = C_4H_6 + C_4H_{10}$$

   2.1.2. Alicyclics

$$2C_6H_8 = C_6H_{10} + C_6H_6$$

   2.2. Disproportionation. Removal of a group and transfer to another molecule.
      2.2.1. Inorganic

$$2CO = C + CO_2 \qquad \text{(Bouduard reaction)}$$

   2.2.2. Aliphatics

$$2C_3H_7CH_3 = C_3H_8 + C_5H_{12}$$

   2.2.3. Aromatics

$$2C_6H_5CH_3 = C_6H_6 + C_6H_4(CH_3)_2$$

   2.3. Polymerization. Formation of long chain molecule, M_n, from nM monomers.
      2.3.1. Aliphatics

Alkene: $nC_3H_6 = $ polypropylene

Alkadiene:

$$nC_4H_4 = \text{rubberlike polymers}$$

   2.3.2. Aromatics

$$nC_6H_5CH{=}CH_2 = \text{polystyrene}$$

   2.3.3. Mixed

two monomer species = copolymer

3. Multimolecular reactions
   3.1. Hydrogenation. Rupture of H–H bond and the addition of H.

   3.1.1. Aliphatics
   Alkene: $C_2H_4 + H_2 = C_2H_6$

   Alkyne: $C_2H_2 + H_2 = C_2H_4$

   Alkadiene: $CH_2CHCHCH_2 + H_2 = C_4H_{10}$

   3.1.2. Alicyclics

   $C_6H_{11}CH_3 + H_2 = C_7H_{16}$ (decyclization)

   3.1.3. Aromatics

   $C_6H_6 + 3H_2 = C_6H_{12}$ (dearomatization)

   $C_6H_6 + 4H_2 = C_6H_{14}$ (dearomatization)

   3.1.4. Oxyorganics
   Aldehyde:

   $CH_3CHO + H_2 = CH_3CH_2OH$ (alkyl)

   $C_6H_5CHO + H_2 = C_6H_5CH_2OH$ (aryl)

   Ester:

   $CH_3COOC_3H_7 + 2H_2 = CH_3CH_2OH + C_3H_7OH$ (alkyl)

   Ketone:

   $CH_3COCH_3 + H_2 = CH_3CHOHCH_3$ (alkyl)

   Acid:

   $HCOOH + 2H_2 = CH_3OH + H_2O$

   Phenol:

   $C_6H_5OH + H_2 = C_6H_6 + H_2O$

   3.1.5. Carbon monoxide

   $CO + H_2 = HCHO$ (formaldehyde synthesis)

   $CO + 2H_2 = CH_3OH$ (methanol synthesis)

   $CO + 3H_2 = CH_4 + H_2O$ (methanation)

   $2CO + 2H_2 = CH_4 + CO_2$ (methanation)

   $nCO + mH_2 = C_nH2_{m-n} + nH_2O$ (Fischer–Tropsch)

   $CO + H_2 + C_3H_6 = C_3H_7CHO$ (hydroformylation)

3.1.6. Nitrogen

$$N_2 + 3H_2 = 2NH_3 \qquad \text{(ammonia synthesis)}$$

3.1.7. Sulfur dioxide

$$SO_2 + 3H_2 = H_2S + 2H_2O$$

3.2. Hydrogenolysis. Rupture of a X–Y bond by H_2.

3.2.1. Rupture of a C–C bond

$$C_2H_6 + H_2 = 2CH_4 \qquad \text{(hydrocracking)}$$

3.2.2. Rupture of a C—O bond
Inorganic:

$$CO + 3H_2 = CH_4 + H_2O \qquad \text{(methanation)}$$

$$3CO + 7H_2 = C_3H_8 + 3H_2O \qquad \text{(Fischer–Tropsch)}$$

$$3CO + 6H_2 = C_3H_6 + 3H_2O \qquad \text{(Fischer–Tropsch)}$$

$$3CO + 6H_2 = C_3H_7OH + 2H_2O \qquad \text{(Fischer–Tropsch)}$$

Aliphatic:

$$CH_3CHO + 2H_2 = C_2H_6 + H_2O$$

$$HCOOH + 2H_2 = CH_3OH + H_2O$$

Aromatic:

$$C_6H_5OH + H_2 = C_6H_6 + H_2O$$

3.2.3. Rupture of a C–S bond
Alkanethiols:

$$C_2H_5SH + H_2 = C_2H_6 + H_2S \qquad \text{(hydrodesulfurization)}$$

Alkyl sulfides:

$$(C_2H_5)_2S + 2H_2 = 2C_2H_6 + H_2S \qquad \text{(hydrodesulfurization)}$$

Thiophenes:

$$C_4H_4S + 4H_2 = C_4H_{10} + H_2S \qquad \text{(hydrodesulfurzation)}$$

3.2.4. Rupture of a C–N bond

$$C_9H_7N + nH_2 = NH_3 + \text{mixed products}$$

$$\text{(hydrodenitrogenation)}$$

3.2.5. Rupture of a N–O bond
Aliphatic:

$$CH_3NO_2 + 3H_2 = CH_3NH_2 + 2H_2O$$

Aromatic:

$$C_6H_5NO_2 + 3H_2 = C_6H_5NH_2 + 2H_2O$$

3.3. Oxidation. The transfer of oxygen with the ruptue of O–O or O–H bonds.

3.3.1. Oxygen addition (direct addition of oxygen)
Aliphatic:

$$2C_2H_4 + O_2 = 2C_2H_4O$$

Aromatic:

$$2C_6H_6 + O_2 = 2C_6H_5OH$$

Aldehyde:

$$2CH_3CHO + O_2 = 2CH_3COOH$$

Inorganic:

$$2CO + O_2 = 2CO_2$$
$$2SO_2 + O_2 = 2SO_3$$

3.3.2. Oxygenolysis (C–H, N–H, S–H rupture with H_2O formation)
Alkane:

$$2C_3H_8 + O_2 = 2C_3H_6 + 2H_2O$$

Alkene:

$$C_4H_8 + O_2 = C_3H_5CHO + H_2O$$
$$C_4H_8 + O_2 = CH_2:CHCOCH_3 + H_2O$$

Alcohol:

$$2C_2H_5OH + O_2 = 2CH_3CHO + 2H_2O$$

Inorganic:

$$4NH_3 + 5O_2 = 4NO + 6H_2O$$
$$2H_2S + O_2 = 2S + 2H_2O$$

3.3.3. Dual oxygenolysis (rupture of C–C and C–H to give CO_2 and H_2O)
Partial oxidation:

$$2C_{10}H_8 + 9O_2 = 2C_6H_4(CO)_2O + 4H_2O$$

Complete combustion:

$$2C_2H_6 + 7O_2 = 4CO_2 + 6H_2O$$

3.3.4. Hydrogenolysis of water (rupture of O–H bond and the formation of H_2)

Inorganic:

$$CO + H_2O = CO_2 + H_2 \qquad \text{(water gas shift)}$$

$$P_4 + 16H_2O = 4H_3PO_4 + 10H_2$$

3.4. Hydration. The transfer of H_2O or active hydrogen without involving gaseous hydrogen.

3.4.1. Intramolecular

Ether:

$$(CH_2H_5)_2O + H_2O = 2C_2H_5OH$$

Alkene:

$$CH_2:CH_2 + H_2O = C_2H_5OH$$

Alkyne:

$$C_2H_2 + H_2O = CH_3CHO$$

Diolefin:

$$CH_2:C:CH_2 + 2H_2O = CH_2OHCH_2CH_2OH$$

3.4.2. Intermolecular

Diene:

$$CH_2:CH(CH_3):CH_2 + H_2O = CH_2:CH_2 + (CH_3)_2CO$$

Aromatic:

$$C_6H_5CH_3 + H_2O = CH_3OH + C_6H_6$$

Ester:

$$HCOOCH_3 + H_2O = CH_3OH + HCOOH$$

Ketone:

$$RCH:CHCHCOCH_3 + H_2O = RCHO + CH_3COCH_3$$

3.5. Amination. The transfer of NH_3.

Inorganic:

$$CO + NH_3 = HCN + H_2O$$

Alcohol:

$$CH_3OH + NH_3 = CH_3NH_2 + H_2O$$

Acid:

$$CH_3COOH + NH_3 = CH_2CONH_2 + H_2O$$

$$CH_3CH_2COOH + NH_3 = CH_3CH_2CN + 2H_2O$$

3.6. Halogenation. The gain of halogen or halide.

    3.6.1. Halogen addition (X-X rupture with no hydrogen halide formation)

    Alkene:

$$CH_2:CH_2 + Cl_2 = ClCH_2CH_2Cl$$

    Alkyne:

$$C_2H_2 + Cl_2 = ClCH:CHCl$$

$$C_2H_2 + 2Cl_2 = CHCl_2CHCl_2$$

    3.6.2. Hydrogen halide addition (H-X rupture)

    Alkene:

$$CH_2:CH_2 + HCl = CH_3CH_2Cl$$

    Alkyne:

$$C_2H_2 + HCl = CH_2:CHCl$$

    Alkyl halide:

$$CH_2:CHCl + HCl = CH_3CHCl_2$$

    3.6.3. By halogen (X-X rupture with formation of hydrogen halide)

    Alkane:

$$CH_4 + Cl_2 = CHCl + HCl$$

$$CH_4 + 2Cl_2 = CH_2Cl_2 + 2HCl$$

$$CH_4 + 4Cl_2 = CCl_4 + 4HCl$$

    Alkene:

$$CH_2:CH_2 + Cl_2 = CH_2:CHCl + HCl$$

    Aromatic:

$$C_6H_6 + Cl_2 = C_6H_5Cl + HCl$$

$$C_6H_6 + 6Cl_2 = C_6Cl_6 + 6HCl$$

    3.6.4. By halogen halide (H-X rupture)

    Aliphatic:

$$C_2H_5OH + HCl = C_2H_5Cl + H_2O$$

    Aromatic:

$$C_6H_5OH + HCl = C_6H_5Cl + H_2O$$

3.6.5. By alkyl halide (C–X rupture with formation of hydrogen halide)
Aromatic:

$$C_6H_6 + C_2H_5Cl = C_6H_5C_2H_5 + HCl$$

3.7. Alkylation. The replacement of an organic hydrogen by an an alkyl radical.
Aliphatic:

$$C_4H_{10} + 2C_4H_8 = C_8H_{18}$$

Aromatic:

$$C_6H_5CH_3 + C_2H_4 = C_6H_4(CH_3)C_2H_5$$
$$C_6H_5C_2H_5 + C_2H4 = C_6H_5C_4H_9$$

3.8. Intermolecular rearrangement.
Aliphatic:

$$C_7H_{16} + C_4H_{10} = C_5H_{12} + C_6H_{14}$$
$$2C_4H_{10} + C_3H_6 = C_8H_{18} + C_3H_8$$

Alicyclic:

$$C_6H_{12} + 3C_2H_4 = C_6H_6 + 3C_2H_6$$

4. Isotopic exchange
4.1. Deuterium. The rupture of a H–H, D–D, H–D, X–H, or X–D bond.
Hydrogen:

$$H_2 + D_2 = 2HD$$

Alkane:

$$CH_4 + CD_4 = 2CH_2D_2$$

Alkene:

$$CH_2:CHD + CH_2:CHCH_3 = CH_2:\ CH_2 + CH_3CH:CHD$$

Inorganic:

$$NH_3 + ND_3 = HN_2D + HND_2$$

4.2. Oxygen

$$M_xO_y^{16} + H_2O^{18} = M_xO_y^{18} + H_2O^{16}$$

4.3. Nitrogen

$$N_2^{14} + N_2^{15} = 2N^{14}N^{15}$$

4.4. Carbon

$$CHC^{14}H_2OH + CO + 3H_2 = CH_3C^{14}H_2CH_3 + 2H_2O$$

APPENDIX 6
Laboratory Recipes

1. SINGLE OXIDES

1.1. Alumina Gel (A Support and Dehydration Catalyst)

Dissolve 1000 g of CP $AlCl_3$ $6H_2O$ in 10,000 cm^3 of H_2O. Prepare an aqueous ammonia solution by adding 840 cm^3 of concentrated ammonia hydroxide (about 30% NH_3) to 1450 cm^3 of H_2O. Add the ammoniacal solution to the $AlCl_3$ solution until the pH is about 8. Allow the precipitate to settle and filter with a Buchner funnel. Wash five times by decantation with 5 cm^3 of concentrated ammonia in 5000 cm^3 of H_2O. Allow the precipitate to settle and filter with a Buchner funnel. By the end of the fifth wash, the precipitate is very difficult to settle since it has started to peptize. If a water wash containing no ammonia is used, peptization begins even during the first wash.

Dry the washed filtrate in an oven for several hours at 100–150°C. Then calcine overnight at about 550°C in a muffle furnace.

1.2. Silica Gel (A Support)

Prepare two solutions. Solution A is 500 cm^3 of water glass ($SiO_2/Na_2O = 3.22$) in 1000 cm^3 of H_2O. Solution B is 245 cm^3 of 4 N HCl plus 300 cm^3 of H_2O. Cool both solutions to about 5°C, then add solution A to solution B with vigorous mechanical agitation. Pour into a flat tray.

The resulting sol has a pH of about 6 and a gel time of 2 min. The gel will be stiff enough to cut into cubes within 30–60 min. Transfer the cubes to a Buchner funnel as soon as possible. Treat immediately with 1 N HCl for a 2-hr period. Repeat this procedure three times using just enough of the HCl solution to completely cover the gel each time.

The gel is washed free of chloride ions and then dried for 4 hr at about 150°C. The product should be transparent at this stage. If any opacity is observed, the drying period should be prolonged. The dried solid is then calcined in air at 550°C.

The calcined product has extremely small pores of about 1 nm, characteristic of dessicant gels. The pore size can be controlled over a wide range by changing the pH of the treating solution. For example, using 1% NH_4Cl instead of 1 N HCl gives a much lower surface area and density, and the pore size will be approximately three times as large.

1.3. Chromic Oxide Gel (Alkane Dehydrogenation)

Add 2500 cm^3 of 0.1 N NH_4 solution at 6 cm^3 per minute to 5000 cm^3 of 0.1 N $Cr(NO_3)_3 \cdot 9H_2O$ at 20°C with stirring. The original precipitate disappears after about 2 hr, but a permanent precipitate forms later. At this point, add the rest of the alkali rapidly. Dark green hydroxide separates as a gelatinous precipitate. Allow the precipitate to settle and decant the supernatant liquid as completely as possible.

Wash the gel with 5% ammonium carbonate solution, adding sufficient solution to bring the level up to the original mark. Stir the suspension for 30 min, allow the precipitate to settle, and decant again. Repeat five times or until the precipitate begins to peptize.

Filter with a Buchner funnel and dry at 100–105°C for 24 hr. Before use, the catalyst is heated in hydrogen while raising the temperature slowly to about 380°C.

2. DUAL OXIDES

2.1. Silica-Alumina Gel (Support and Cracking Catalyst)

Prepare two solutions. Solution A is 465 cm^3 of water glass in 900 cm^3 of H_2O. Solution B is 67 cm^3 of 4 N HCl plus 140 cm^3 of 0.5 N $Al_2(SO_4)_3$ solution in 400 cm^3 of H_2O. Cool both solutions to about 5°C. Add solution A to solution B rapidly with vigorous stirring. Pour into a tray as soon as mixing is complete. The gel time is about 35 s. In about 1 hr, cut the gel into cubes approximately 2 cm on a side. Allow the gel to age for 48 hr and transfer to a Buchner funnel. Base exchange with a 2% solution of $Al_2(SO_4)_3 \cdot 18H_2O$ three times for 2-hr periods and then once overnight. Wash with distilled water until free of sulfate ions. Dry for 4 hr at 140 to 180°C.

Continue the drying until the gel pieces are translucent. Calcine 10-15 hr at 550°C.

2.2. Nickel Oxide Alumina (Hydrogenation and Methanation)

Dissolve 454 g of $Al(NO_3)_3 9H_2O$ in 3000 cm³ of H_2O and cool to 5-10°C. Dissolve 200 g of NaOH in 1000 cm³ of H_2O and cool to 5-10°C. Add the soldium hydroxide solution to the aluminum nitrate solution dropwise while stirring vigorously over a period of 1-2 hr. Dissolve 100 g of $Ni(NO_3)_2 6H_2O$ in 600 cm³ of H_2O, add 45 cm³ of concentrated HNO and cool to 5-10 °C. Add the nickel nitrate solutions to the sodium aluminate solution with vigorous stirring over a period of 0.5-1 hr. Filter the light green precipitate in a Buchner funnel and suspend the filtrate in 2000 cm³ of H_2O for 15 min while stirring. Repeat the washing procedure six times. Cut the filtrate into cubes, place in an evaporating dish, and dry in an air oven for 16 hr at 105°C. Crush the catalyst to 8-12 mesh.

Before use, the required amount of catalyst is reduced in a stream of hydrogen (30-80 cm³ min^{-1}) for 16 hr at 350°C. The furnace temperature is raised to the reduction temperature over a period of 3 hr.

3. DISPERSED OXIDES

3.1. Chromia Alumina (Dehydrogenation)

Place 175 g of activated alumina of the desired mesh size in a 500-cm³ filter flask. Attach a dropping funnel through a one-hole stopper. At some point between the stopcock of the funnel and the rubber stopper, make a flexible tubing connection that may be clamped shut. Attach the side arm of the flask to a vacuum pump. Dissolve 31.6 g of CrO_2 in sufficient water to make 65 cm³ of solution and place the solution in the dropping funnel. Evacuate the flask and admit the chromium solution while shaking the catalyst to ensure uniform absorption of the liquid. When all the solution has been added and the catalyst is uniformly impregnated, remove the catalyst from the flask, dry at 150°C, and calcine in hydrogen at 550°C.

3.2. Fischer–Tropsch (Hydrocarbon Synthesis)

Dissolve 246 g of $Co(NO_3)_2 6H_2$ and 6.3 g of $Th(NO_3)_4 4H_2O$ in 1300 cm³ of H_2O. Heat to boiling and add 100 g of kieselguhr and 6 g of MgO in 500 cm³ of H_2O, maintaining the boiling. Dissolve 92 g of sodium carbonate in 500 cm³ of H_2O and heat to boiling. Add the boiling slurry of kieselguhr and MgO to the carbonate and nitrate solutions simultaneously, stirring vigorously. Filter and wash with distilled water until free of sodium ions. Dry the filter cake in an oven at 110°C overnight. Crush to 8–12 mesh.

Heat the catalyst in a furnance in a stream of hydrogen, reaching 400°C in 2 hr. Hold at 400°C for 2 hr.

3.3. Cobalt-Moly Alumina (Hydrodesulfurization)

Prepare a solution of ammonium paramolybdate by dissolving 92 g in 300 cm³ of H_2O. Add this solution to 525 g of dehydrated thete-alumina in the form of 0.8 mm extrudates. Allow the mixture to stand, with agitation, for 30 min. Remove the extrudates and dry for 4 hr at 110°C. Prepare a cobalt solution by dissolving 85 g of $Co(NO_3)_2 6H_2O$ in 250 cm³ of H_2O. Add this solution to the dried extrudates and soak for 30 min. Then dry at 110°C for 10 hr. Calcine at 550°C for 10 hr, increasing the temperature at the rate of 1°C per minute.

The catalyst is activated by sulfiding in hydrogen containing 10% hydrogen sulfide at 400°C for 4 hr.

APPENDIX 7
Catalyst Manufacturers

1. CATALYST CODES

1.1. Petroleum refining
Catalytic cracking	P1
Hydrotreating	P2
Hydrocracking	P4
Reforming	P4

1.2. Chemical processing
Hydrogenation	C1
Dehydrogenation	C2
Polymerization	C3
Oxidation	C4
Ammonia synthesis	C5
Methanol synthesis	C6
Steam reforming	C7

1.3. Emission control
Automobile exhaust	A1
Stack control	A2

1.4. Carriers and supports
Alumina	S1
Silica	S2
Other supports	S3

2. CATALYST MANUFACTURERS IN THE UNITED STATES

Company	Product
Activated Metals, Sevierville, TN	C1
Air Products and Chemicals, Allentown, PA	P2, P4, C1, C2, C3

263

Alcoa Chemicals, Pittsburgh, PA	S1
American American Cyanamid, Wayne, NJ	P2, P3, P4, C3, C4
BASF Wyandotte, Parsippany, NJ	C1, C4
Calsicat, Erie, PA	C1, C3
Catalyst Resources, Houston, TX	C3
Davison, Baltimore, MD	P1, P2, P3, C1, C3, C4, A1
Degussa, S. Plainfield, NJ	P1, P4, C1, C2, C3, C4, A1
Ethyl, Baton Rouge, LA	C3
Foote Mineral, Exton, PA	C3
Halcon SD, Little Ferry, NJ	C4
Haldor Topsoe, Houston, TX	P2, C4, C5, C6, C7
Harshaw/Filtrol, Cleveland, OH	P1, P2, P3, C1, C4
Johnson Matthey, Deptford, NJ	C1, A1
Kaiser Chemicals, Oakland, CA	S1
Katalco, Oak Brook, IL	P2, C1, C5, C6, C7
Katalistiks, Baltimore, MD	P1
Ketjen Catalysts, Pasadena, TX	P1, P2
Lithium Corp., Gastonia, NC	C3
Lucidol, Buffalo, NY	C3
Monsanto, St. Louis, MO	C4
Norton, Akron, OH	S1, S2
Noury, Chicago, IL	C3
Platina, S. Plainfield, NJ	C1, C2, C3, C4
PQ Corp., Valley Forge, PA	S1
SCM Metals Products, Cleveland, OH	C4
Shell Chemical, Houston, TX	P2, P3, C1, C2, C4
Stauffer Chemical, Westport, Con	C3, C4
Texas Alkyls, Westport, Con	C3
UOP, Des Plains, IL	P2, P3, P4, C1,F4C2, A1
Union Carbide, Danbury, Con	C3, C4
United Catalysts, Louisville, KY	C1, C2, C4, C5, C6, C7
U.S. Peroxygen, Richmond, CA	C3
Vista, Houston, TX	S1

3. CATALYST MANUFACTURERS IN WESTERN EUROPE

Company	Product
Achima, Switzerland	C3
Akzo Chemie, Netherlands	P1, C1, C3, C4
AUSINP Industries, Italy	C4
BASF, West Germany	P2, P3, C1, C2, C4, C5, C6, A1

Bayer, West Germany	C1, C4
CLA, France	C4
Crosfield, UK	C1, C2
Cynamid-Ketjen, Netherlands, UK	P2, P4
Degussa, West Germany	C1, C3, C4, A1
Doduco, West Germany	C1, C4, A1
Dutral, Italy	C1, C4
Dynamit Nobel, West Germany	C3
Engelhard, UK	C1, C4
Engelhard Kalichemie, West Germany	P4, A1
Ethyl, Belgium	C3
Exocat, France	A1
Frauenthal/SGP, Austria	A1
FBC Ind. Chemical, UK	C3
FSOS, France	C3
Grace Feldmuhle, West Germany	P1, A1
Haldor Topsoe, Denmark	C1, C4, C5, C6, C7, A1
Harshaw/Filtrol, Netherlands	C1, C2
Haraeus, West Germany	A1
Hoechst, West Germany	C1
Huls/Sud Chemie, West Germany	C1, A1
ICI Ltd, UK	C1, C5, C6, C7
Interox, West Germany, Belgium	C3
Johnson Matthey, UK	C1, C4, A1
Kali-Chemie, West Germany	P1, A1
Katec, West Germany	A1
Katalistiks, Netherlands	P1
Luperox (Lucidol), West Germany	C3
Lihco, UK	C3
M&T Chemicals, UK	C3
NL Kronos, Belgium, West Germany	C3
Norton, UK	A1
Peroxide Catalyst, UK	C3
Pennwalt, Netherlands	C3
Procatalyse, France	P2, P4, C1
Schering, West Germany	C3
Servo, Netherlands	C3
Shell Chemical, Belgium, UK	P2, C1, C2, C4
Than et Mulhouse, France	C3
Tioxide, UK	C3
UCI/SUD Chemie, Belgium, West Germany	C1, C2, C4
Unikat, Sweden	A1
Unichema, West Germany	c1
UOP, UK	P3, P4, C1, A1
Wacker, West Germany	C3

4. CATALYST MANUFACTURERS IN JAPAN

Company	Product
Catalysts and Chemicals, Far East, Tokyo	P4, C1, C2, A1
Catalysts and Chemical Industries, Tokyo	P1, P2, P3, C1, C4, A1
Kawaken Fine Chemicals, Tokyo	C1
Mitsui Bussan Chemical Machinery, Tokyo	C3
NGK Insulators, Nagoya City	A1
Nikki Chemical, Tokyo	C1, C2, C4, A1
Nikki Universal, Tokyo	P2, P4, C2, A1
Nippon Engelhard, Tokyo	C1, C4, A1
Nippon Ketjen, Tokyo	P2
Nippon Shokubai Kagaku Kogyo, Osaka	C4, A1
Nissan Girdler, Tokyo	P2, C1, C2, A1
Tanaka Kikinzoku Kogyo, Tokyo	C1, C4, A1
Toyo Stauffer, Tokyo	C3

REFERENCES

1. C. Vancini, *Synthesis of Ammonia*, Macmillan, London (1971).
2. I. R. Shannon, The Synthesis of Ammonia and Related Reactions, *Catalysis*, Vol. 2 (C. Kimbal and D. A. Dowden, eds.) p. 28, The Chemical Society, London (1978).
3. J. M. Thomas and W. J. Thomas, *Introduction to the Principles of Heterogeneous Catalysis*, Academic, New York (1967).
4. G. A. Mills and F. W. Steffgen, Catalytic Methanation, *Catal. Rev.* **8**, 159 (1973).
5. R. B. Anderson, Nitrided Iron Catalysts for the Fischer–Tropsch Synthesis in the Eighties, *Catal. Rev.* **21**, 53 (1980).
6. E. K. Poels and V. Ponec, Formation of Oxygenated Products from Synthesis Gas, *Catalysis*, Vol. 6 (G. C. Bond and G. Webb, eds.) p. 196, Royal Society of Chemistry, London (1983).
7. F. Marschner and F W. Moeller, Methanol Synthesis, *Applied Industrial Catalysis*, Vol. 2 (B. E. Leach, ed.) p. 215, Academic, New York (1983).
8. R. Hughes, *Deactivation of Catalysts*, Academic, New York (1984).
9. H. Heinmann, Catalytic Routes to 100 Octane Gasoline—1939 to 1980, *Proceedings of the Robert A. Welch Foundation XXV. Heterogeneous Catalysis*, The Robert A. Welch Foundation, Houston (1983).
10. F. S. Kovarik and J.B. Butt, Reactor Optimization in the Presence of Catalytic Decay, *Catal. Rev.* **24**, 441 (1982).
11. J. H. Sinfelt, *Bimetallic Catalysts, Discoveries, Concepts, and Applications*, John Wiley, New York (1983).
12. M. Orchin, Homogeneous Catalysis: A Wedding of Theory and Experiment, *Catal. Rev.*, **26**, 59 (1984).
13. S. A. Topham, The History of the Catalytic Synthesis of Ammonia, *Catalysis, Science and Technology*, Vol. 7 (J. R. Anderson and M. Boudart, ed.), p. 1, Springer-Verlag, New York (1985).
14. G. Ertl, Surface Science and Catalysis—Studies on the Mechanism of Ammonia Synthesis, *Catal. Rev.* **21**, 201 (1980).
15. J. B. McKinley, The Hydrodesulfurization of Liquid Petroleum Fractions, *Catalysis*, Vol. 5 (p. Emmett, ed.), p. 405, Reinhold, New York (1957).
16. P. C. H. Mitchell and C. E. Scott, Outstanding Problems of Hydrodesulfurization Catalysis, *Bull. Soc. Chim. Belges* **93**, 619 (1984).
17. G. W. Parshall, *Homogeneous Catalysis*, John Wiley, New York (1980).

18. G. P. Royer, Immobilized Enzymes Catalysis, *Catal. Rev.* **22**, 29 (1980).
19. E. G. Christoffel, Laboratory Reactors and Heterogeneous Catalytic Processes, *Catal. Rev.* **24**, 159 (1982).
20. H. L. Lee, *Heterogeneous Reactor Design*, Butterworths, New York (1984).
21. C. N. Satterfield, *Mass Transfer in Heterogeneous Catalysis*, MIT Press, Boston (1970).
22. J. F. LePage and J. Miguel, Determining Mechanical Properties of Industrial Catalysts, *Preparation of Catalysts I* (B. Delmon, P.A. Jacobs and G. Poncelet, eds.), p. 39, Elsevier, Amsterdam (1976),
23. A. Clark, *The Theory of Adsorption and Catalysis*, Academic, New York (1970).
24. G. C. Bond, *Catalysis by Metals*, Academic, New York (1962).
25. J. H. DeBoer, Adsorption Phenomena, *Advances in Catalysis*, Vol. 8 (W. G. Frankenburg, V. I. Komarewsky and E. K. Rideall, eds.), p. 17, Academic, New York (1956).
26. I. Toyoshima and G. A. Somorjai, Heats and Adsorption of O_2, H_2, CO, CO_2 and N_2 on Polycrystalline and Single Crystal Transition Metal Surfaces, *Catal. Rev.* **19**, 105 (1979).
27. J. J. Carberry, *Chemical and Catalytic Reaction Engineering*, McGraw-Hill, New York (1976).
28. M. Boudart, *Kinetics of Chemical Processes*, Prentice-Hall, Englewood Cliffs, New Jersey (1968).
29. K.J. Laidler, Kinetic Laws in Surface Catalysis, *Catalysis*, Vol. 1 (P. H. Emmet, ed.), p. 119, Reinhold, New York (1954).
30. M. Boudart, Kinetics and Mechanism of Ammonia Synthesis, *Catal. Rev.* **23**, 1 (1981).
31. J. M. Smith, *Chemical Engineering Kinetics*, McGraw-Hill, New York (1980).
32. A. Wheeler, Reaction Rates and Selectivity in Catalyst Pores, *Advances in Catalysis*, Vol. 3, p. 250, Academic, New York (1950).
33. E. E. Gonzo and J. C. Gottifredi, Rational Approximations of Effectiveness Factors and General Diagnostic Criteria for Heat and Mass Transfer Limitations, *Catal. Rev.* **25**, 119 (1983).
34. J. W. Fulton, Selecting the Catalyst Configuration, *Chem. Eng.* **May 12**, 97 (1986).
35. A. V. Stiles, *Catalyst Manufacture*, Marcel Dekker, New York (1983).
36. M. W. Ranney, *Desulfurization of Petroleum*, Noyes Data, Park Ridge, New Jersey (1975).
37. J. R. H. Ross, Metal Catalysed Methanation and Steam Reforming, *Catalysis*, Vol. 7 (G. C. Bond and G. Webb, eds.) p. 1, The Royal Society of Chemistry, London (1985).
38. J. R. Rostrup-Nielsen, Catalytic Steam Reforming, *Catalysis, Science and Technology*, Vol. 5 (J. R. Anderson and M. Boudart, eds.), p. 1, Springer-Verlag, New York (1984).
39. D. C. Puxley, I. J. Kitchener, C. Komodromos, and N. D. Parkyns, The Effect of Preparation Method Upon the Structure, Stability and Metal/Support Interactions in Nickel/Alumina Catalysts, *Preparation of Catalysts III* (G. Poncelet, P. Grange, and P. A. Jacobs, eds.), p. 237, Elsevier, Amsterdam (1983).
40. C. H. Bartholomew and J. R. Katzer, Sulfur Poisoning of Nickel in CO Hydrogenation, *Catalyst Deactivation* (B. Delmonand G. F. Froment, eds.), p. 375, Elsevier, Amsterdam (1981).
41. C. H. Bartholomew, Carbon Deposition in Steam Reforming and Methanation, *Catal. Rev.* **24**, 67 (1982).
42. D. L. Trimm, The Formation and Removal of Coke from Nickel Catalysts, *Catal. Rev.* **16**, 155 (1977).
43. W. B. Innes, Catalyst Carriers, Promoters, Accelerators, Poisons and Inhibitors, *Catalysis*, Vol. 1 (P. B. Emmett, ed.), p. 245, Reinhold, New York (1954).
44. G. C. Bond, *Catalysis by Metals*, Academic, New York (1962).
45. O. V. Krylov, *Catalysis by Non-Metals*, Academic, New York (1970).

46. K. Tanabe, Solid Acid and Base Catalysis, *Catalysis, Science and Technology*, Vol. 2, (J. R. Anderson and M. Boudart, eds.), p. 231, Springer-Verlag, New York (1981).

47. K. Taylor, Automobile Catalytic Converters, *Catalysis, Science and Technology*, Vol. 5 (J. R. Anderson and M. Boudart, eds.), p. 119, Springer-Verlag, New York (1984).

48. H. N. Holmes, *Laboratory Manual of Colloid Chemistry*, John Wiley, New York (1928).

49. S. P. S. Andrew, *Catalyst Handbook*, p. 20, Springer-Verlag, New York (1970).

50. H. H. Lee and Ruckenstein, Catalyst Sintering and Reactor Design, *Catal. Rev.* 25, 475 (1983).

51. F. Moseley, R. W. Stephens, K. D. Stewart, and J. Wood, The Poisoning of a Steam Hydrocarbon Gasification Catalyst, *J. Catal.* 24, 18 (1972).

52. R. A. Rajadhyaksha and L. K. Doraiswamy, Falsification of Kinetic Parameters by Transport Limitations, *Catal. Rev.* 13, 209 (1976).

53. C. Mista (ed.), *Industrial Alumina Chemicals*, ACS Monograph Series No. 184, American Chemical Society, Washington, D. C. (1986).

54. R. Iler, *The Chemistry of Silica*, Wiley, New York (1979).

55. K. Tanabe, *Solid Acids and Bases*, Academic, New York (1970).

56. F. G. Ciapetta, R. M. Dobres, and R. W. Baker, Catalytic Reforming of Pure Hydrocarbons and Petroleum Naphthas, *Catalysis*, Vol. 6 (P. H. Emmett, ed.), p. 497, Reinhold, New York (1957).

57. J. W. Ward, Design and Preparation of Hydrotreating Catalysts, *Preparation of Catalysts III* (G. Poncelet, P. Grange, and P. A. Jacobs, eds.), p. 587, Elsevier, Amsterdam (1983).

58. W. F. Taylor, D. J. C. Yates, and J. H. Sinfelt, Catalysis Over Supported Metals: The Effect of the Support on the Catalytic Activity of Nickel for Ethane Hydrogenolysis, *J. Phys. Chem.* 68, 2962 (1964).

59. A. T. Bell, Supports and Metal-Support Interactions in Catalyst Design, *Catalyst Design, Progress and Perspectives* (L. L. Hegedus, ed.), p. 103, Wiley, New York (1987).

60. R. Van Hardeveld and F. Hartog, Influence of Metal Particle Size in Nickel-on-Aerosil Catalysts on Surface Site Distribution, Catalytic Activity, and Selectivity, *Advances in Catalysis*, Vol. 22 (D. D. Eley, H. Pines, and P. B. Weisz, eds.), p. 75, Academic, New York (1972).

61. C. P. Haung and J. T. Richardson, The Activity of Nickel on Sodium-Neutralized Silica-Alumina, *J. Catal.* 52, 332 (1978).

62. P. A. Sermon and G. C. Bond, Hydrogen Spillover, *Catal. Rev.* 8, 211 (1973).

63. G. C. Bond and R. Burch, Strong Metal-Support Interactions, *Catalysis*, Vol. 6, (G. C. Bond and G. Webb, eds.) p. 27, Royal Society of Chemistry, London (1983).

64. R. K. Oberlander, Aluminas for Catalysts—Their Preparation and Properties, *Applied Industrial Catalysis*, Vol. 3 (B. E. Leach, ed.) p. 63, Academic, New York (1984).

65. R. Gauguin, M. Graulier, and D. Papee, Thermally Stable Carriers, *Adv. Chem. Ser.* 143, 147 (1975).

66. W. D. Mross, Alkali Doping in Heterogeneous Catalysis *Catal. Rev.* 25, 591 (1982).

67. G. A. Somorjai, The Building of Catalysts: A Molecular Surface Science Approach, *Catalyst Design, Progress and Perspectives* (L. L. Hegedus, ed.), p. 11, Wiley, New York (1987).

68. W. G. Frankenburg, The Catalytic Synthesis of Ammonia from Hydrogen and Nitrogen, *Catalysis*, Vol. 3 (P. H. Emmett, ed.), p. 171, Reinhold, New York (1955).

69. J. G. Speight, *The Desulfurization of Heavy Oils and Residua*, Marcel Dekker, New York (1981).

70. D. C. McCulloch, Catalytic Hydrotreating in Petroleum Refining, *Applied Industrial Catalysis*, Vol. 1 (B. E. Leach, ed.), p. 70, Academic, New York (1983).

71. J. T. Richardson, Magnetic Study of Cobalt Molybdenum Oxide Catalysts, *I&ED Fundamentals* 3, 154 (1964).

72. R. Rahnasamy and S. Sivasanker, Structural Chemistry of Co–Mo–Alumina Catalysts, *Catal. Rev.* **22**, 401 (1980).

73. H. Topsoe and B. S. Clansen, Importance of Co–Mo–S Type Structures in Hydrodesulfurization, *Catal. Rev.* **26**, 395 (1984).

74. J. T. Richardson, Electronic Properties of Unsupported Cobalt-Promoted Molybdenum Sulfide, *J. Catal.* **112**, 313 (1988).

75. P. C. H. Mitchell, Sulfide Catalysts, Characterization and Reactions Including Hydrodesulfurization, *Catalysis*, Vol. 4 (C. Kemball and D. A. Dowden, ed.), p. 175, Royal Society of Chemistry, London (1981).

76. D. Chadwick, D. W. Aitchinson, R. Badilla-Ohlbaum, and L. Josefsson, Influence of Phosphorus on the HDS activity of Ni–Mo/Al$_2$O$_3$ Catalysts, Preparation of Catalysts III (G. Poncelet, P. Grange, and P. A. Jacobs, eds.), p. 323, Elsevier, Amsterdam (1983).

77. K. Onuma, Preparation of Bimodal Alumina and Other Inorganic Oxides Suitable for Hydrotreating Catalysts, Preparation of Catalysts IV (B. Delmon, P. Grange, P. A. Jacobs, and P. Poncelet, eds.), p. 543, Elsevier, Amsterdam (1987).

78. D. G. Gavin, Coal Hydrogenation Catalysts, *Catalysis*, Vol. 5 (G. C. Bond and G. Webb, eds.), p. 220, The Chemical Society, London (1982).

79. F. Aerstin and G. Street, *Applied Chemical Process Design*, Plenum Press, New York (1978).

80. J. H. Gary and G. L. Handwerk, *Petroleum Refining, Technology and Economics*, Marcel Dekker, New York (1975).

81. F. A. Holland, F. A. Watson, and J. K. Wilkinson, *Introduction to Process Economics*, John Wiley, New York (1984).

82. B. E. Leach, Industrial Catalysis: Chemistry Applied to Your Life-Stile and Environment, *Applied Industrial Catalysis*, Vol. 1 (B. E. Leach, ed.), p. 1, Academic, New York (1981).

83. G. A. Somorjai, *Chemistry of Two Dimensions: Surfaces*, Cornell University Press, Ithica (1981).

84. G. F. Froment, Catalytic Kinetics: Modelling, *Catalysis Science and Technology*, Vol. 2 (J. R. Anderson and M. Boudart, eds.), p. 97, Springer-Verlag, New York (1981).

85. D. L. Trimm, *Design of Industrial Catalysts*, Elsevier, New York (1980).

86. J. M. Berty, Laboratory Reactors for Catalytic Studies, *Applied Industrial Catalysis*, Vol. 1 (B. E. Leach, ed.), p. 43, Academic, New York (1983).

87. J. Horak and J. Pasek, *Design of Industrial Chemical Reactors from Laboratory Data*, Heyden, London (1978).

88. M. P. Dudukovic and P. L. Mills (eds.), *Chemical Catalytic Reactor Modeling*, ACS Symposium Series No. 237, American Chemical Society (1983).

89. Y. Ogino, Catalysis by Molten Metals and Molten Alloys, *Catal. Rev.* **23**, 505 (1981).

90. M. P. Rosynek, Catalytic Properties of Rare Earth Oxides, *Catal. Rev.* **16**, 111 (1977).

91. R. J. Madix, Reaction Kinetics and Mechanisms: Model Studies on Metal Single Crystals, *Catal. Rev.* **26**, 281 (1984).

92. J. Haber, Crystallography of Catalyst Types, *Catalysis, Science and Technology*, Vol. 2 (J. R. Anderson and M. Boudart, eds.,), p. 13, Springer-Verlag, New York (1981).

93. S. J. Thomson, Catalysis on Well-Defined Metal Surfaces and Non-Metallic Surfaces, *Catalysis*, Vol. 1 (C. Kemball, ed.), p. 1, The Chemical Society, London (1977).

94. R. W. Joyner, Catalysis by Single Crystal Surfaces, *catalysis*, Vol. 5 (G. C. Bond and G. Webb, eds.), p. 1, Royal Society of Chemistry, London (1982).

95. G. Ertl, Kinetics of Chemical Processes on Well-Defined Surfaces, *Catalysis, Science and Techno.ogy*, Vol. 4 (J. R. Anderson and M. Boudart, eds.), p. 209, Springer-Verlag, New York (1985).

96. A. A. Balandin, Modern State of the Multiplet Theory of Heterogeneous Catalysis, *Advances in Catalysis*, Vol. 19, p. 1, Academic, New York (1969).

97. M. Boudart and G. Djega-Mariadasson, *Kinetics of Heterogeneous Catalytic Reactions*, Princeton University Press, Princeton, New Jersey (1984).

98. C. Kittel, *Introduction to Solid State Physics*, John Wiley, New York (1953).

99. H. M. Hulburt, The Nature of Catalytic Surfaces, *Catalysis*, Vol. 2 (P. H. Emmett, ed.), p. 167, Reinhold, New York (1955).

100. T. S. Cale and J. T. Richardson, In-Situ Characterization of Ni-Cu/SiO$_2$ Catalysts, *J. Catal.* **79**, 378 (1983).

101. P. G. Menon and T. S. R. Prasada, Surface Enrichment in Catalysis, *Catal. Rev.* **20**, 97–120 (1979).

102. R. L. Moss, Reactions of Hydrocarbons on Alloy and Bimetallic Catalysts, *Catalysis*, Vol. 1 (C. Kemball, ed.), p. 37, The Chemical Society, London (1977).

103. D. A. Dowden, Reactions of Hydrocarbons on Multimetallic Catalysis, *Catalysis*, Vol. 2 (C. Kemball and D. A. Dowden, eds.), p. 1, The Chemical Society, London (1978).

104. W. M. H. Sachtler, Chemisorption on Alloy Surfaces, *Catal. Rev.* **14**, 193 (1976).

105. J. B. Goodenough, *Magnetism and the Chemical Bond*, p. 57, Interscience Publishers, New York (1963).

106. G. C. Bond, Adsorption and Co-ordination of Unsaturated Hydrocarbons with Metal Surfaces and Metal Atoms, *Disc. Farad. Soc.* **41**, 200 (1966).

107. N. D. Spencer, R. C. Schoonmaker, and G. A. Somorjai, Iron Single Crystals as Ammonia Synthesis Catalysts, *J. Catal.* **74**, 129 (1982).

108. R. Van Hardeveld and F. Hartog, Influence of Metal Particle Size in Nickel-on-Aerosil Catalysts on Surface Site Distribution, Catalytic Activity and Selectivity, *Advances in Catalysis*, Vol. 22 (D. D. Eley, H. Pines, and P. B. Weiss, eds.), p. 75, Academic, New York (1972).

109. G. C. Bond, Adsorptive and Catalytic Properties of Small Metallic Crystallites, *Fourth International Congress on Catalysis* (J. W. Hightower, ed.), Rice University, Houston, p. 1227 (1968).

110. P. H. Desai and J. T. Richardson, Crystallite Size Effects in Nickel Catalysts: Cyclohexane Dehydrogenation and Hydrogenolysis, *J. Catal.* **98**, 392 (1986).

111. G. A. Ozin, Very Small Metallic and Bimetallic Clusters, *Catal. Rev.* **16**, 191 (1977).

112. R. F. Marzke, Quantum Size Effects in Small Metallic Particles, *Catal. Rev.* **19**, 43 (1979).

113. S. D. Jackson, P. B. Wells, R. Whynman, and P. Worthington, Metal Clusters and Cluster Catalysts, *Catalysis*, Vol. 4 (C. Kemball and D. A. Dowden, eds.), p. 75, Royal Society of Chemistry, London (1981).

114. E. L. Muetterties, Molecular Clusters as Catalysts, *Catal. Rev.* **23**, 69 (1981).

115. G. L. C. Maire and F. G. Garin, Metal Catalyzed Skeletal Reactions of Hydrocarbons on Metal Catalysts, *Catalysis, Science and Technology*, Vol. 6, (J. R. Anderson and M. Boudart, eds.), p. 161, Springer-Verlag, New York (1984).

116. J. C. Vickerman, Catalytic Properties of Oxide Solid Solutions, *Catalysis*, Vol. 2 (C. Kemball and D. A. Dowden, eds.), p. 107, Royal Society of Chemistry, London (1978).

117. F. S. Stone, Catalytic Properties of Oxides, *Chemistry of the Solid State* (W. E. Garner, ed.), Butterworths, London (1955).

118. F. F. Vol'kenstein, The Electronic Theory of Catalysis on Semiconductors, *Advances in Catalysis*, Vol. 12 (D. D. Eley, P. W. Selwood and P. B. Weisz, eds.), p. 189, Academic, New York (1960).

119. F. F. Vol'kenstein, *The Electronic Theory of Catalysis on Semiconductors*, Macmillan, New York (1963).

120. F. F. Vol'kenstein, The Electronic Theory of Catalysis on Semiconductors, *Problems Kinetics Catal.* **8**, 79 (1955).

121. F. F. Vol'kenstein, The Electronic Theory of Photocatalytic Reactions on Semiconductors, *Advances in Catalysis*, Vol. 23, p. 157 (D. D. Eley, H. Pines, and P. B. Weisz, eds.), Academic, New York (1973).

122. M. Formenti and S. J. Teichner, Heterogeneous Photocatalysis, *Catalysis*, Vol. 2 (C. Kemball and D. A. Dowden, eds.), p. 87, The Chemical Society, London (1978).

123. R. I. Bickley, Heterogeneous Photocatalysis, *Catalysis*, Vol. 5 (G. C. Bond and G. Webbs, eds.), p. 308, Royal Society of Chemistry, London (1982).

124. R. I. Bickley, T. Gonzalez-Carreno, and L. Palmisano, the Preparation and the Characterization of Some Ternary Titanium Oxide Photocatalysts, *Preparation of Catalysts IV* (B. Delmon, P. Grange, P. A. Jacobs, and G. Poncelet, eds.), p. 297, Elsevier, Amsterdam (1987).

125. P. Mark, A Comparison of Chemical Activity at Ordered and Disordered Semiconductor Surfaces, *Catal. Rev.* **12**, 71 (1975).

126. C. S. John and M. S. Scurrell, Catalytic Properties of Aluminas for Reactions of Hydrocarbons and Alcohols, *Catalysis*, Vol. 1 (C. Kemball, ed.), p. 136, The Chemical Society, London (1977).

127. J. B. Peri, A Model for the Surface of γ-alumina, *J. Phys. Chem.* **69**, 220 (1965).

128. M. W. Tamele, Chemistry of the Surface and the Activity of Alumina-Silica Cracking Catalyst, *Disc. Faraday Soc.* **8**, 270 (1950).

129. H. Heinemann, A Brief History of Industrial Catalysis, *Catalysis, Science and Technology*, Vol. 1 (J. R. Anderson and M. Boudart, eds.), p. 1, Springer-Verlag, New York (1981).

130. V. Haensel, Catalytic Cracking of Pure Hydrocarbons, *Advance in Catalysis*, Vol. 3 (W. G. Frankenburg, V. I. Komarewsky, and E. K. Rideal, eds.), p. 179, Academic, New York (1951).

131. A. G. Oblad, T. H. Milliken, Jr., and G. D. Mills, Chemical Characteristics and Structure of Cracking Catalysts, *Advances in Catalysis*, Vol. 3 (W. G. Frankenburg, V. I. Komarewsdky, and E. K. Rideal, eds.), p. 199, Academic Press, New York (1951).

132. J. A. Rabo, Unifying Principles in Zeolite Chemistry and Catalysis, *Catal. Rev.* **23**, 293 (1981).

133. E. M. Flanigen, Molecular Sieve Materials: Their Synthesis, Properties and Characteristics, *Catal. Rev.* **26**, 483 (1984).

134. R. Rudham and A. Stockwell, Catalysis on Faujasitic Zeolites, *Catalysis*, Vol. 1 (C. Kemball, ed.), p. 87, The Chemical Society, London (1977).

135. H. Heinemann, Technological Applications of Zeolites in Catalysis, *Catal. Rev.* **23**, 315 (1981).

136. J. W. Wood, Molecular Sieve Catalysts, *Applied Industrial Catalysis*, Vol. 3 (B. E. Leach, ed.), p. 272, Academic, New York (1984).

137. W. O. Haag and N. Y. Chen, Catalysts Design with Zeolites, *Catalyst Design, Progress and Perspective* (L. L. Hegedus, ed.), p. 163, John Wiley, New York (1987).

138. B. Imelik, C. Naccache, Y. Ben Taarit, J. C. Vedrine, G. Coudurier, and H. Praliand, (ed.), *Catalysis by Zeolites*, Elsevier, Amsterdam (1980).

139. W. M. Meier and J. B. Uytterhoeven (eds.), *Molecular Sieves*, Elsevier, Amsterdam (1980).

140. C. D. Chang, Hydrocarbons from Methanol, *Catal. Rev.* **25**, 1 (1983).

141. J. T. Richardson, The Effect of Faujasite Cations on Acid Sites, *J. Catal.* **9**, 182 (1967).

142. T. E. Whyte, Jr., and R. A. Dalla Betta, Zeolite Advances in the Chemical and Fuel Industries: A Technical Perspective, *Catal. Rev.* **24**, 567 (1982).

143. P. A. Jacobs, Acid Zeolites: An Attempt to Develop Unifying Concepts, *Catal. Rev.* **24**, 415 (1982).

144. T. Takeshita, R. Ohnishi, and K. Tanabe, Recent Survey of Catalysis by Solid Metal Sulfates, *Catal. Rev.* **8**, 29 (1973).

145. J. B. Moffat, Phosphates as Catalysts, *Catal. Rev.* **18**, 199 (1978).

146. M. A. M. Boersma, Catalytic Properties of Alkali Metal-Graphite Compounds, *Catal. Rev.* **10**, 243 (1974).

147. S. Malinowski and J. Kijenski, Superbasic Heterogeneous Catalysts, *Catalysis*, Vo.. 4 (C. Kemball and D. A. Dowden, eds.), p. 130, Royal Society of Chemistry, London (1981).

148. S. T. Oyama and G. L. Heller, Catalysis by Carbides, Nitrides and Group VIII Intermetallic Compounds, *Catalysis*, Vol. 5 (G. C. Bond and G. Webb, eds.), p. 333, Royal Society of Chemistry, London (1982).

149. D. A. Dowden, C. R. Schnell, and G. T. Walker, The Design of Complex Catalysts, *Fourth International Congress on Catalysis* (J. W. Hightower, ed.), Rice University, Houston, p. 1120 (1968).

150. Z. Paal and P. Tetenge, Reactions of Hydrocarbons on Metallic Catalysts, *Catalysis*, Vol. 5 (G. C. Bond and G. Webb, eds.), p. 80, Royal Society of Chemistry, London (1982).

151. A. Bielouski and J. Haber, Oxygen in Catalysis on Transition Metal Oxides, *Catal.* **19**, 1 (1979).

152. G. K. Boreskov, The Catalyis of Isotopic Exchange in Molecular Oxygen, *Advances in Catalysis*, Vol. 15, p. 285, Academic, New York (1964).

153. G. K. Boreskov, Mechanism of Catalytic Oxidations Reactions on Solid Catalysts, *Kin. Kat.* **14**, 1 (1873).

154. M. Raney, Catalysis from Alloys, *Ind. Eng. Chem.* **32**, 1199 (1940).

155. H. B. Weiser, *The Hydrous Oxides*, McGraw-Hill, New York (1976).

156. D. J. Shaw, *Introduction to Colloid and Surface Chemistry*, Butterworths, London (1980).

157. G. J. K. Acres, A. J. Bird, J. W. Jenkins, and F. King, The Design and Preparation of Supported Catalysts, *Catalysis*, Vol. 4 (C. Kemball and D. A. Dowden, eds.), p. 1, Royal Society of Chemistry, London (1981).

158. R. K. Iler, *The Chemistry of Silica*, John Wiley, New York (1979).

159. M. E. Winyall, Silica Gels: Preparation and Properties, *Applied Industrial Catalysis*, Vol. 3 (B. E. Leach, ed.), p. 43, Academic, New York (1984).

160. N. Pernicone and F. Traina, Commercial Catalysts Preparation, *Applied Industrial Catalysis*, Vol. 3 (B. E. Leach, ed.), p. 1, Academic, New York (1984).

161. J. W. Fulton, Making the Catalyst, *Chem. Eng.*, July 7, 59 (1986).

162. J. T. Richardson, unpublished results.

163. M. Sittig, *Handbook of Catalysts Manufacture*, Noyes Data, Park Ridge, New Jersey (1978).

164. P. C. Puxley, I. J. Kitchener, C. Komodromos, and N. D. Parkyns, The Effect of Preparation Method Upon the Structures, Stability and Metal/Support Interactions in Nickel/Alumina Catalysts, *Preparation of Catalysts III* (G. Poncelet, P. Grange, and P. A. Jacobs, eds.), p. 237, Elsevier, Amsterdam (1983).

165. H. B. Weiser and W. O. Milligan, The Mechanism of the Coagulation of Sols by Electrolytes, *J. Phys. Chem.* **40**, 1071 (1936).

166. C. E. Hofstadt, M. Schneider, O. Bock, and K. Kochloefl, Effect of Preparation Methods of Cu-ZnO-A$_2$O$_3$-K Catalysts, *Preparation of Catalysts III* (G. Poncelet, P. Grange, and P. A. Jacobs, eds.), p. 709, Elsevier, Amsterdam (1983).

167. A. Culfaz and L. B. Sand, Mechanism of Nucleation and Crystallization of Zeolites from Gels, *Molecular Sieves* (W. M. Meier and J. B. Uytterhoeven, eds.), Advances in Chemistry Series 121, American Chemical Society, Washington, D.C. (1973).

168. L. A. M. Hermans and J. W. Geus, Interaction of Nickel Ions with Silica Supports During Deposition—Precipitation, *Preparation of Catalysts II* (B. Delmon, P. Grange, P. Jacobs, and G. Poncelet, eds.), p. 113, Elsevier, Amsterdam (1979).

169. C. J. Pereira, G. Kim, and L. L. Hegedus, A Novel Catalyst Geometry for Automobile Emission Control, *Catal. Rev.* **26**, 503 (1984).

170. J. T. Richardson, unpublished results.

171. H. P. Stephens and R. G. Dosch, Catalyst Preparation via Hydrous Metal Oxide Ion-Exchangers, *Preparation of Catalysts IV* (B. Delmon, P. Grange, P. A. Jacobs, and J. Poncelet, eds.), p. 271, Elsevier, Amsterdam (1987).

172. M. Komiyama, Design and Preparation of Impregnated Catalysts, *Catal. Rev.* 27, 341 (1985).

173. M. Kotter and L. Riekert, The Influence of Impregnation, Drying and Activation on the Activity and Distribution of CuO on α-Alumina, *Preparation of Catalysts II* (B. Delmon, P. Grange, P. Jacobs, and G. Poncelet, eds.), p. Elsevier, Amsterdam (1979).

174. S. Y. Lee and R. Aris, The Distribution of Active Ingredients in Supported Catalysts Prepared by Impregnation, *Catal. Rev.* 27, 207 (1981).

175. J. L. G. Fierro, P. Grange, and B. Delmon, Control of Concentration Profiles by Rational Preparation of Pelleted Hydrodesulfurization Catalysts, *Preparation of Catalysts IV* (B. Delmon, P. Grange, P. Jacobs, and G. Poncelet, eds.), p. 591, Elsevier, Amsterdam (1987).

176. J. T. Richardson, unpublished results.

177. V. V. Boldyrev, M. Buleus, and B. Delmon, *The Control of the Reactivity of Solids*, p. 87, Elsevier, Amsterdam (1979).

178. J. T. Richardson and R. J. Dubus, Preparation Variables in Nickel Catalysts, *J. Catal.* 54, 207 (1978).

179. R. Prada Silvy, J. L. G. Fierro, P. Grange, and B. Delmon, Influence of the Activation Procedure on the Nature and Concentration of the Active Phase in Hydrodesulfurization Catalysts, *Preparation of Catalysts IV* (B. Delmon, P. Grange, P. Jacobs, and G. Poncelet, eds.), p. 605, Elsevier, Amsterdam (1987).

180. E. Lieber and F. L. Morritz, The Uses of Raney Nickel, *Advances in Catalysis*, Vol. 5 (W. G. Frankenburg, E. K. Rideal, and V. I. Komarewsky, eds.), p. 417, Academic, New York (1953).

181. Ph. Courty and Ch. Marcilly, A Scientific Approach to the Preparation of Bulk Mixed Oxide Catalysts, *Preparation of Catalysts III* (G. Poncelet, P. Grange, and P. Jacobs, eds.), p. 485, Elsevier, Amsterdam (1983).

182. S. P. S. Andrew, Heterogeneous Catalyst Preparation: The Fabrication of Microstructures, *Preparation of Catalysts I* (B. Delmon, P. A. Jacobs, and G. Poncelet, eds.), p. 429, Elsevier, Amsterdam (1976).

183. C. L. Thomas, *Catalytic Processes and Proven Catalysis*, Academic, New York (1970).

184. W. S. Briggs, Catalysts and the Automobile—Twenty-Five Years Later, *Applied Industrial Catalysis*, Vol. 3 (B. E. Leach, ed.), p. 241, Academic, New York (1984).

185. K. Taylor, Automobile Catalysis Converters, *Catalysis, Science and Technology*, Vol. 5 (J. R. Anderson and M. Boudart, eds.), p. 119, Springer-Verlag, New York (1984).

186. J. M. Winton, Catalysts '88 Restructuring for Technical Clout, *Chemical Week* **June 29**, 20 (1988).

187. E. F. Sanders and E. J. Schlossmacher, Catalyst Scale-Up—Pitfall or Payoff? *Applied Industrial Catalysis*, Vol. 1 (B. E. Leach, ed.), p. 31, Academic, New York (1983).

188. K. Jiratova, L. Janacek, and P. Schneider, Influence of Aluminum Hydroxide Peptization on Physical Properties of Alumina Extrudates, *Preparation of Catalysts III* (G. Poncelet, P. Grange, and P. A. Jacobs, eds.), p. 653, Elsevier, Amsterdam (1983).

189. R. M. Cahen, J. M. Andre, and H. R. Debus, Process for the Production of Spherical Catalyst Supports, *Preparation of Catalysts II* (B. Delmon, P. Grange, P. Jacobs, and G. Poncelet, eds.), p. 585, Elsevier, Amsterdam (1979).

190. R. O. Feuge, Hydrogenation of Glyceride Oils, *Catalysis*, Vol. 3 (P. H. Emmett, ed.), p. 413, Reinhold, New York (1955).

191. F. Delannay and B. Delmon, Methods of Catalyst Characterization: An Overview, *Characterization of Heterogeneous Catalysis* (F. Delannay, ed.), p. 1, Marcel Dekker, New York (1984).

192. A. H. Neal, Organization and Functions of ASTM Committee D-32 on Catalysts, *Preparation of Catalysts II* (B. Delmon, P. Grange, P. Jacobs, and G. Poncelet, eds.), p. 719, Elsevier, Amsterdam (1979).

193. ASTM Committee D-32, *1985 Annual Book of ASTM Standards*, Vol. 5.03, American Society for the Testing of Materials, New York (1984).

194. E. G. Derouane, The Council of Europe Research Group on Catalysts, *Preparation of Catalysts II* (B. Delmon, P. Grange, P. Jacobs, and G. Poncelet, eds.), p. 727, Elsevier, Amsterdam (1979).

195. G. K. Boreskov, Measurement of the Activity of Solid State Catalysts, *Preparation of Catalysis II* (B. Delmon, P. Grange, P. Jacobs, and G. Poncelet, eds.), p. 723, Elsevier, Amsterdam (1979).

196. T. Hattori, H. Matsumoto, and Y. Murakami, Standardization of Catalyst Test Methods by the Committee on Reference Catalysts of the Catalysis Society of Japan, *Preparation of Catalysis IV* (B. Delmon, P. Grange, P. A. Jacobs, and G. Poncelet, eds.), p. 815, Elsevier, Amsterdam (1987).

197. B. D. Cullity, *Elements of X-Ray Diffraction*, Addison-Wesley, Reading, Massachusetts (1956).

198. S. J. B. Reed, *Electron Microprobe Analysis*, Cambridge University Press, Cambridge (1975).

199. B. Welz, *Atomic Absorption Spectroscopy*, Verlag Chemie, Weinheim (1976).

200. F. Delannay, *Analytical Electron Microscopy of Heterogeneous Catalysis, Catal. Rev.* **22**, 141 (1980).

201. P. Gallezot, X-Ray Techniques in Catalysis, *Catalysis, Science and Technology*, Vol. 5 (J. R. Anderson and M. Boudart, eds.), p. 221, Springer-Verlag, New York (1984).

202. J. V. Sanders, The Electron Microscopy of Catalysis, *Catalysis, Science and Technology*, Vol. 7 (J. R. Anderson and M. Boudart, eds.), p. 5, Springer-Verlag, New York (1985).

203. T. Baird, Characterization of Catalysts by Electron Microscopy, *Catalysis*, Bol. 5 (G. C. Bond and G. Webbs, eds.), p. 172, Royal Society of London, London (1982).

204. F. Delannay, Transmission Electron Microscopy and Related Microanalytical Technics, *Characterization of Heterogeneous Catalysts* (F. Delannay, eds.), p. 71, Marcel Deller, New York (1984).

205. C. J. Wright, Catalyst Characterization with Neutron Techniques, *Catalysis*, Vol. 7 (G. C. Bond and G. Webbs, eds.), p. 46, Royal Society of London, London (1985).

206. W. W. Wendlandt, *Thermal Methods of Analysis*, Interscience, New York (1964).

207. T. Daniels, *Thermal Analysis*, Kogan Page, London (1973).

208. A. Jones and B. D. McNicol, Temperature Programmed Reduction, *Catal. Rev.* **24**, 233 (1982).

209. J. L. Lemaitre, Temperature-Programmed Methods, *Characterization of Heterogeneous Catalysts* (F. Delannay, ed.), p. 29, Marcel Dekker, New York (1984).

210. J. R. Anderson and K. C. Pratt, *Introduction to Characterization and Testing of Catalysts*, Academic Press, New York (1985).

211. T. Allen, *Particle Size Measurement*, Chapman and Hall, London (1975).

212. C. Orr and J. M. Dallavalle, *Fine Particle Measurement*, Macmillan, New York (1959).

213. P. C. Carmen, *Flow of Gases Through Porous Media*, Butterworths, London (1956).

214. J. F. Page and J. Miguel, Determining Mechanical Properties of Industrial Catalysts: Correlations with Their Morphological and Physical-Chemical Properties, *Preparation of Catalysts I* (B. Delmon, P. A. Jacobs, and G. Poncelet, eds.), p. 39, Elsevier, Amsterdam (1976).

215. P. E. Emmett, Measurement of the Surface Area of Solid Catalysts, *Catalysis*, Vol. 1 (P. E. Emmett, ed.), p. 31, Reinhold, New York (1954).

216. A. J. Lecloux, Texture of Catalysts, *Catalysis, Science and Technology*, Vol. 2 (J. R. Anderson and M. Boudart, eds.), p. 171, Springer-Verlag, New York (1981).

217. J. C. P. Broekhoff, Mesopore Determination from Nitrogen Sorption Isotherms: Fundamentals, Scope, Limitations, *Preparation of Catalysts II* (B. Delmon, P. Grange, P. Jacobs, and G. Poncelet, eds.), p. 633, Elsevier, Amsterdam (1979).

218. J. T. Richardson, Experimental Determination of Catalyst Fouling Parameters, *I & EC Proc. Res. Dev.* **11**, 12 (1972).

219. W. N. Delgass, G. Haller, R. Kellerman, and J. H. Lunsford, *Spectroscopy in Heterogeneous Catalysis*, Academic, New York (1979).

220. J. M. Thomas and R. M. Lambert (eds.), *Characterization of Catalysts*, John Wiley, New York (1980).

221. J. I. Goldstein and H. Yakowitz (eds.), *Practical Scanning Electron Microscopy*, Plenum Press, New York (1975).

222. R. T. K. Baker, In Situ Electron Microscopy Studies of Catalyst Particle Behavior, *Catal. Rev.* **19**, 161 (1979).

223. J. L. Lemaitre, P. G. Menon, and F. Delannay, The Measurement of Catalyst Dispersion, *Characterization of Heterogeneous Catalysis* (F. Delannay, ed.), p. 229, Marcel Dekker, New York (1984).

224. P. Ganesan, H. Kud, A. Saavedra, and R. J. de Angelis, Particle Size Distribution Function of Supported Metal Catalysts by X-Ray Diffraction, *J. Catal.* **52**, 310 (1978).

225. T. A. Carlson, *Photoelectron and Auger Spectroscopy*, Plenum Press, New York (1974).

226. C. Defosse, X-Ray Photoelectron Spectroscopy, *Characterization of Heterogeneous Catalysts* (F. Delannay, ed.), p. 225, Marcel Dekker, New York (1984).

227. R. I. Declerck-Grimee, P. Canesson, R. M. Friedman, and J. J. Fripiat, An X-Ray Photoelectron Spectroscopy Study of Various $CoMo/Al_2O_3$ Hydrodesulfurization Catalysts, *J. Phys. Chem.* **82**, 889 (1978).

228. J. T. Richardson and T. S. Cale, Interpretation of Hydrogen Chemisorption on Nickel Catalysts, *J. Catal.* **102**, 419 (1986).

229. J. H. Sinfelt and D. J. C. Yates, Catalytic Hydrogenolysis of Ethane Over Noble Metals of Group VIII, *J. Catal.* **8**, 82 (1967).

230. J. R. Anderson, *Structure of Metallic Catalysts*, Academic, New York (1975).

231. J. H. Sinfelt, W. F. Taylor, and D. J. C. Yates, Catalysis Over Supported Metals III, comparison of Metals of Known Surface Area for Ethane Hydrogenolysis, *J. Phys. Chem.* **69**, 95 (1965).

232. T. Fukushima, S. Galvagno, and G. Parravano, Oxygen Chemisorption on Supported Gold, *J. Catal.* **57**, 177 (1979).

233. K. Taylor, Determination of Ruthenium Surface Areas by Hydrogen and Oxygen Chemisorption, *J. Catal.* **38**, 299 (1975).

234. P. H. Emmett, S. Brunauer, Accumulation of Alkali Promoters on Surfaces of Iron Synthetic Ammonia Catalysts, *J. Am. Chem. Soc.* **59**, 310 (1937).

235. P. H. Emmett and N. Skan, The Catalytic Hydrogenation of Benzene Over Metal Catalysts, *J. Am. Chem. Soc.* **65**, 1029 (1943).

236. J. J. F. Scholten and A. Van Montfoort, The Determination of the Free-Metal Surface Area of Palladium Catalysts, *J. Catal.* **1**, 85 (1962).

237. J. L. G. Fierro, S. Mendioroz, J. A. Pajares, and S. W. Weller, Specific Surface Area of Molybdena in MoO_3/SiO_2 Catalysts, *J. Catal.* **65**, 263 (1980).

238. K. Otto and M. Shelef, The Adsorption of Nitric Oxide on chromia Supported on Alumina, *J. Catal.* **14**, 226 (1969).

239. H. S. Gandhi and M. Shelef, The Adsorption of Nitric Oxide and Carbon Monoxide on Nickel Oxide, *J. Catal.* **24**, 241 (1972).

240. K. Otto and M. Shelef, The Adsorption of Nitric Oxide on Iron Oxides, *J. Catal.* **18**, 184 (1970).

241. C. R. F. Lund, J. J. Schorfheide, and J. A. Dumesic, Magnetite Surface Area Titration Using Nitric Oxide, *J. Catal.* **57**, 105 (1979).

242. J. J. F. Scholten, Metal Surface Area and Metal Dispersion in Catalysis, *Preparation of Catalysis II* (B. Delmon, P. Grange, P. Jacobs, and G. Poncelet, eds.), p. 685, Elsevier, Amsterdam (1979).

243. J. E. Benson and M. Boudart, Hydrogen–Oxygen Titration Method for the Measurement of Supported Platinum Surface Area, *J. Catal.* **4**, 704 (1965).

244. J. J. F. Scholten and J. A. Konvalinka, Reaction of Nitrous Oxide with Copper Surfaces, *Trans. Faraday Soc.* **65**, 2465 (1969).

245. J. J. F. Scholten, J. A. Konvalinka, and F. W. Beekman, Reaction of Nitrous Oxide and Oxygen with Silver Surfaces, and Application to the Determination of Free-Silver Surface Area of Catalysts, *J. Catal.* **28**, 209 (1973).

246. J. T. Richardson, Dual-Functional Activities in Nickel Acid Catalysts, *J. Catal.* **21**, 122 (1971).

247. T. R. Hughes and H. M. White, A Study of the Surface Structure of Decationized Y Zeolite by Quantitative Infrared Spectroscopy, *J. Phys. Chem.* **71**, 2192 (1967).

248. P. E. Eberly, J., High-Temperature Infrared Spectroscopy of Pyridine Adsorbed on Faujasites, *J. Phys. Chem.* **72**, 1042 (1968).

249. H. V. Drushel and A. L. Sommers, Catalyst Acidity Distribution Using Visible and Fluorescent Indicators, *Anal. Chem.* **38**, 1723 (1966).

250. E. V. Ballou, R. T. Barth, and R. A. Flinn, *J. Phys. Chem.* **65**, 1639 (1961).

251. A. K. Galwey, Compensation Effect in Heterogeneous Catalysis, *Advances in Catalysis*, Vol. 26 (D. D. Eley, H. Pines, and P. B. Weisz, eds.), p. 247, Academic, New York (1977).

252. L. K. Doraiswamy and D. G. Tajbl, Laboratory Catalytic Reactors, *Catal. Rev.* **10**, 177 (1974).

253. E. G. Christoffel, Laboratory Reactors and Heterogeneous Catalytic Processes, *Catal. Rev.* **24**, 149 (1982).

254. J. M. Berty, Testing Commercial Catalysts in Recycle, Reactors, *Catal. Rev.* **20**, 75 (1979).

255. T. Furusawa, M. Suzuki, and J. M. Smith, Rate Parameters in Heterogeneous Catalysts by Pulse Techniques, *Catal. Rev.* **13**, 43 (1976).

256. G. A. Somorjai, Active Sites in Heterogeneous Catalysis, *Advances in Catalysis*, Vol. 26 (D. D. Eley, H. Pines, and P. B. Weisz, eds.), p. 2, Academic, New York (1977).

257. P. J. Denny and M. V. Twigg, Factors Determining the Life of Industrial Heterogeneous Catalysts, *Catalyst Deactivation* (B. Delmon and G. F. Froment, eds.), p. 577, Elsevier, Amsterdam (1980).

258. J. B. Butt, Catalyst Deactivation and Regeneration, *Catalysis, Science and Technology*, Vol. 6 (J. R. Anderson and M. Boudart, eds.), p. 1, Springer-Verlag, New York (1984).

259. F. S. Kovarik and J. B. Butt, Reactor Optimization in the Presence of Catalyst Decay, *Catal. Rev.* **24**, 441 (1982).

260. J. T. Richardon, SNG Catalyst Technology, *Hydrocarbon Processing*, December 1973, p. 91.

261. P. B. Venuto and E. T. Habib, Jr., *Fluid Catalytic Cracking with Zeolite Catalysts*, Marcel Dekker, New York (1979).

262. W. M. H. Sachtler and R. A. van Santen, Surface Composition and Selectivity of Alloy Catalysts, *Advances in Catalysis*, Vol. 26 (D. D. Eley, H. Pines, and P. B. Weisz, eds.), p. 69, Academic Press, New York (1977).

263. B. Delmon and P. Grange, Solid State Chemical Phenomena in Ageing and Deactivation of Catalysts, *Catalyst Deactivation* (B. Delmon and G. F. Froment, eds.), p. 507, Elsevier, Amsterdam (1980).

264. G. C. Kuczynski, A. E. Miller, and G. A. Sargent (Editors), *Sintering and Heterogeneous Catalysis*, Plenum Press, New York (1984).

265. R. Hughes, *Deactivation of Catalysts*, Academic, New York (1984).

266. J. T. Richardson and J. G. Crump, Crystallite Size Distributions of Sintered Nickel Catalysts, *J. Catal.* **57**, 417 (1979).

267. J. T. Richardson and R. Koveal, Influence of Crystallite Size on Carbon Monoxide Methanation, *J. Catal.* **98**, 559 (1986).

268. J. T. Richardson and J. L. Propp, Pore Size Effects on Sintering of Ni/Al_2O_3 Catalysts, *J. Catal.* **98**, 457 (1986).

269. P. Desai and J. T. Richardson, Support Effects During Sintering of Nickel Catalysts, *Catalyst Deactivation* (B. Delmon and G. F. Froment, eds.), p. 149, Elsevier, Amsterdam (1981).

270. S. F. Adler and J. J. Kearney, The Physical Nature of Supported Platinum, *J. Phys. Chem.* **64**, 208 (1960).

271. H. J. Maat and L. Moscou, A Study of the Influence of Platinum Crystallite Size on the Selectivity of Platinum Reforming Catalysts, *Third International Congress on Catalysis*, p. 1277, North-Holland, Amsterdam (1965).

272. J. Barbier, Effect of Poisons on the Activity of Selectivity of Metallic Catalysts, *Deactivation and Poisoning of Catalysts* (J. Oudar and H. Wise, eds.), p. 109, Marcel Dekker, New York (1985).

273. E. B. Maxted, The Poisoning of Metallic Catalysts, *Advances in Catalysis*, Vol. 3 (W. G. Frankenburg, E. K. Rieal, and V. I. Komarewsky, eds.), p. 129, Academic, New York (1951).

274. J. T. Richardson, Sulfiding of Nickel Catalyst Beds, *J. Catal.* **21**, 130 (1971).

275. L. L. Hegedus and R. W. McCabe, Catalyst Poisoning, *Catal. Rev.* **23**, 377 (1981).

276. H. Knozinger, Specific Oising and Characterization of Catalystically Active Oxide Surfaces, *Advances in Catalysis*, Vol. 25 (D. D. Eley, H. Pines, and P. B. Weisz, eds.), p. 184, Academic, New York (1976).

277. J. H. Sinfelt, Catalytic Reforming of Hydrocarbons, *Catalysis, Science and Technology*, Vol. 1 (J. R. Anderson and M. Boudart, eds.), p. 257, Springer-Verlag, New York (1981).

278. E. E. Wolf and F. Alfani, Catalyst Deactivation by Coking, *Catal. Rev.* **24**, 329 (1982).

279. C. Naccache, Deactivation of Acid Catalysts, *Deactivation and Poisoning of Catalysts* (J. Oudar and H. Wise, eds.), p. 185, Marcel Dekker, New York (1985).

280. A. Voorhies, Jr., Carbon Formation in Catalytic Cracking, *Ind. Eng. Chem.* **37**, 318 (1945).

281. D. M. Nace, Catalytic Cracking over Crystalline Aluminosilicates, *Ind. Eng. Chem. Prod. Res. Dev.* **8**, 24 (1969).

282. J. P. Franck and G. P. Martino, Deactivation of Reforming Catalysts, *Deactivation and Poisoning of Catalysts* (J. Oudar and H. Wise, eds.), p. 205, Marcel Dekker, New York (1985).

283. J. R. Rostrup-Nielsen and D. L. Trimm, Mechanisms of Carbon Formation on Nickel-Containing Catalysts, *J. Catal.* **48**, 155 (1977).

INDEX

Abrasion, catalyst, 132
Absorption, effects in catalysis, 7
Acetic acid, 7
Acidity, measurement of, 166
Acids
  catalyzed reactions, 28, 75
  coking on, 210
  dehydration by, 27, 75
  ionization of, 72
  poisoning of, 207
Activation, of the catalyst, 117
Active component
  deposition of, 108
  precipitation of, 109
  role of, 26
Activity, 237
  benefits of, 5
  decay in coking, 213
  dual functional, 34, 36, 76
  measurement of, 17
  practical, 180
  from temperature, 1
Adsorption
  of active component, 111
  chemical, 15, 146
  effects in catalysis, 7
  physical, 15, 146
  transport effects in, 113
Agglomeration, in catalyst preparation, 99
Alchemist
  in catalyst design, 35
  in catalyst preparation, 95
Alkylates, 237
Alkylation, 237

Alumina
  Brönsted sites in, 33, 69, 71
  dehydration of, 33, 71
  effect of calcination on, 105
  effect of impurities on, 192
  hydrous oxides, 33
  Lewis sites in, 33
  loss of activity after regeneration, 193
  metal contamination from porphorins, 219
  phase transitions, 35, 104
  reaction with oxides, 35
  recipe for, 259
Aluminum chloride treating, 237
Ammonia decomposition, 4
Ammonia synthesis
  catalysts for, 50
  effect of temperature and pressure, 3
  energetics of, 1
  history of, 7
  mechanism of, 2
  promoters in, 37
  steam reforming in, 25
  surface effects in, 57
  volcano curve for, 17
Analytical electron spectroscopy, 137
Anions
  adsorption, 112
  effect in preparation, 97, 109
Areal rate, 173
Aromatic cycloparaffins, 237
Aromatization, 237
Arrhenius parameters, as measure of activity, 174

ASTM Committee, characterization standards, 136
ASTM distillation, 237
Atomic absorption spectroscopy, 137
Atomic emission spectroscopy, 137
Attrition
    in catalyst particles, 133
    measurement of, 144
Auger electron spectroscopy (AES), 160
Automobile exhaust catalyst, deactivation in, 189
Average pore radius, nomenclature, 11
Axial crushing strength, 143

Barrel, 237
    per calendar day, 237
    per stream day, 237
Base chemisorption, 171
Battery limits, 237
Berty reactor, 178
BET equation, 147
Bicycloparaffins, 238
Bitumen, 238
Blending, 238
Bottoms, 238
Bromine index, 238
Bromine number, 238
Brönsted sites
    carbon formation on, 36
    nature of, 33
Bulk crushing strength, measurement of, 144
Bulk properties, 136

Caffeine number, 388
Calcination
    effect on acidity of γ-Al_2O_3, 73
    effect on concentration profiles, 117
    effect on reduction of NiO/Al_2O_3, 118
    role of, 104
Carbanion, 238
Carberry reactor, 178
Carbon formation
    on Brönsted sites, 36
    in methanation, 26
    in steam reforming, 26, 221
Carbonium ions, 238
    mechanisms, 28
Catalysis
    compromises in, 23
    electrical conductivity and, 26
    enzyme, 8

Catalysis (cont.)
    heterogeneous, 7
    homogeneous, 6
    organization of, 6
    reference books in, 225
    research, 44
Catalyst
    components, 26
    definition of, 1
    design of, 45
    development of, 41
    dual functional, 34
    engineering, 23, 45
    manufacture of, 123
    novel, 45, 96
    preparation of, 95
    testing, 46
    types, 49
Catalyst design, 45
    by computers, 92
    methodology, 83
Catalyst formulation
    methods for, 127
    pilot unit testing of, 46
Catalyst manufacturers, 263
    in Japan, 265
    in the United States, 263
    in Western Europe, 264
Catalyst particles
    common types, 9
    density, 141
    effect of shape and size, 8, 23, 44
    mechanical properties of, 9, 23
Catalyst properties
    bulk, 136
    composition, 136
    mechanical, 143
    particle, 140
    phase structure, 137
    surface, 157
Catalyst stability, 24
    in methanation, 25
    in steam reforming, 25
Catalytic activity
    importance of, 5
    of insulators and solid acids, 26, 75
    measurement of, 171
    of metals, 26, 62
    of oxides and sulfides, 26, 66
Catalytic cracking, 238
    deactivation in, 189

Catalytic cracking (*cont.*)
  particle size of catalysts in, 142
  poisoning by metals, 191
  selectivity in, 6
  by solid acids, 27, 82
  with ZSM-5, 81
Catalytic reforming, 238
  deactivation in, 189, 215
  dual functional activity in, 34
  effect of crystallite size on selectivity, 199
  mechanism of, 34
  by metals, 50
  promotion by rhenium, 37, 193
  segregation of metals in, 193
Cation adsorption, 112
Cetene number, 238
Chemisorption
  for dispersion measurement, 162
  on metals, 165
  nature of, 146
  on nonmetals, 165
  stoichiometry of, 164
Classification of the active components, 27
Clays
  acid sites in, 73
  as catalysts, 73
Cloud point, 238
Coalescence
  of metal crystallites, 29
  of sols, 100
CoAl$_2$O$_4$, in hydrodesulfurization catalysts, 38
Coal liquefaction
  direct, 43
  indirect, 50
Cobalt, reactions catalyzed by, 62
Coking
  correlations in, 211
  deactivation by, 210
  dehydrogenative, 214
  dissociative, 220
  regeneration after, 210
Commercial manufacturing of catalysts, 123
Complete combustion, 238
CoMo/Al$_2$O$_3$
  analysis of, 137
  coking in, 217
  diffusivity in, 157
  promotion with K, 39
  promotion with P, 39
  recipe for, 262
  shape of particles, 24

CoMo/Al$_2$O$_3$ (*cont.*)
  structure of, 38
  surface structure of, 161
Composition
  solution methods, 136
  spectroscopic methods, 137
Computer
  -aided design of catalysts, 92
  control of pilot unit, 47
Concentration profiles
  during adsorption, 112
  during impregnation, 115
Condensation, 238
Confidentiality, in catalyst manufacturing, 126
Conradson carbon, 238
Contamination coke, 212
Controlled atmosphere electron microscopy
    (CAEM), 159
CoO, 38
  in hydrodesulfurization catalysts, 38
Copolymer, 238
Copper, reactions catalyzed by, 62
Coprecipitation, of mixed hydrous oxides, 107
Cracking, 238, 251
Cr$_2$O$_3$ gel, recipe for, 160
Cr$_2$O$_3$/Al$_2$O$_3$, recipe for, 261
Crystallite size
  effect of reduction, 118
  effect on selectivity, 199
  in metals, 59
  in semiconductors, 68
Cu–Ni alloys
  cyclopropane hydrogenolysis, 55
  phase segregation, 193
Cut, 238
Cu-ZnO-Al$_2$O$_3$
  preparation of, 121
  sintering of, 197
Cu-ZnO-Al$_2$O$_3$, support effects, 31
Cyclization, 238
Cyclohexane dehydrogenation, 61
Cycloparaffins, 238
Cyclopropane hydrogenation, 202

Deactivation
  cause of, 191
  by coking, 210
  by component volatization, 192
  by compound formation, 193
  effects of, 6, 185
  effects on temperature profiles, 190

Arrhenius parameters (cont.)
  by fouling, 192
  by heavy metals, 192
  kinetics of, 188
  by metal contamination, 218
  by poisoning, 200
  process, 187
  role of, 5
  by sintering, 194
Dearomatization, 238
Dehydration, 238
  by solid acids, 27, 75
Dehydrogenation, 238
  coking, 214
  by metals, 214
  pattern for oxides, 90
Demetallization, 239
Denitrogenation, on sulfides, 28
Densities, 141
Design equation, 176
Desulfurization, 239
Diagnostic tests, 21
Differential thermal analysis (DTA), 140
Diffusion
  bulk, 13
  during adsorption, 113
  effects in catalysis, 7
  external, 11
  index of, 20
  interaction with chemical rates, 20
  Knudsen, 13
  measurement of, 155
  surface, 16
  tests for, 21
  of xylenes in ZSM-5, 81
Dinuclear aromatics, 239
Dismutation, 239
Dispersion
  from chemisorption, 164
  definition of, 28
  measurement of, 162
  from poison titration, 166
  from reaction titration, 166
Disproportionation, 239, 252
Dissociative coking, 210, 220
Distillation, 238
Doctor test, 239
Drying
  commercial units for, 103
  concentration profiles and, 116
  after deposition, 111

Drying (cont.)
  of hydrogels, 102
  temperature effects in, 103
Dual functional activity, 34, 36, 76
Dual oxygenolysis, 239, 255

Effectiveness factor, 14, 19
Electron diffraction, 138
Electronic theory, 16, 53
  of metals, 16, 53
  of semiconductors, 68
Electron microscopy, 157
Electron probe analysis, 137
Energy conservation, in catalyst manufacturing, 126
Enzyme catalysts, 8
Equipment, in catalyst manufacturing, 125
Ethane hydrogenation, effect of the support, 34
Extended x-ray absorption fine structure (EXAFS), 160
Extrudates
  manufacture of, 128
  pressure drop with, 10

Faujasite
  acidity of, 168
  H_2S poisoning of, 168
  ion exchange in, 79, 115
  preparation of, 108
  structure of, 77
Fe molybdate, for methane oxidation, 91
Fe_3O_4, for methane oxidation, 91
Fermi level
  of metals, 54
  of semiconductors, 54
Filtration, of hydrogels, 102
Fischer–Tropsch
  reactions, 239
  recipe for catalyst, 261
Fixed carbon, 239
Flakes, manufacture of, 130
Fluidized beds, 189
Fluorescent indicators, 168
Formylation, 239
Fouling
  cause of, 192
  effect of, 192
Free carbon, 239

Gas hourly space velocity (GHSV), 239

Gas oil, 239
Gelation
  of silica sols, 101
  in support preparation, 99
Geometric theory, 16, 52
Granules, manufacture of, 130
Guard beds, 189

Halogenation, 239, 255
Hammett indicators, 168
Heat transfer limited reaction, 239
Heavy metals
  contamination by, 218
  deactivation of, 219
  removal of, 219
Heterogeneous catalysis, 7
Hexane isomerization, poisoning of, 209
Homogeneous catalysis, 6
Huit equation, 143
Hydration, 239, 256
Hydrocarbons, critical diameter of, 79
Hydrocarboxylation, 239
Hydrocracking, 239
  dual functionality in, 34
  with metals, 50
  with zeolites, 81
Hydrodealkylation, 239
Hydrodenitrogenation, 239
Hydrodesulfurization, 239
  catalyst activation in, 120
  catalyst functions in, 37
  catalyst particle shape and size, 24
  deactivation in, 189
  history of, 7
  on sulfides, 28
Hydroformylation, 7, 240
Hydrogels
  preparation of, 99
  processing of, 129
Hydrogen
  adsorption, 58
  effect of flow on reduction, 118
  spillover, 34
  transfer, 240, 252
Hydrogenation, 240, 253
  of cyclopropane, 55
  by metals, 27, 50
  selective, 27
Hydrogenolysis, 240, 254
  by metals, 27, 62
  by oxides and sulfides, 27, 70

Hydrogenolysis (*cont.*)
  site selectivity of, 60
  sulfur poisoning of, 215
Hydrotreating, 217, 240
Hydrous oxides
  alumina, 33
  mixed, 107
  preparation of, 95
  silica, 33

Ignition loss, 145
Impregnation, of the active component, 115, 116
Infra-red spectroscopy, 157
Ink bottle pores, 155
Insulators, in catalysis, 26, 69
Interfacial phenomena, 7
Ion exchange
  in catalyst preparation, 114
  in zeolites, 75
Iron
  porphorin contamination, 219
  reactions catalyzed by, 62
Isomerization, 240, 251
  by insulators, 27
Isotope exchange, 240, 258

Kelvin equation, 152
Kerosine, 240
Kinetics
  activities from, 172
  deactivation, 198
  differential, 175
  first-order rate constants, 174
  integral, 175
  Langmuir–Hinshelwood, 174
Knudsen diffusion, 157

Langmuir–Hinshelwood
  approximation, 174
  rate equation, 17, 174, 176
Langmuir isotherm, 17, 146
Legal responsibilities, for catalysts, 133
Lewis sites, in γ-Al_2O_3, 33
Lifetime, 240
  testing with pilot units, 46
Liquefied petroleum gas (LPG), 240
Liquid hourly space velocity (LHSV), 240

Lobes
  manufacture of, 128
  pressure drop with, 10
Localized bond theory, 56

Macropores, 10
Mass transfer coefficient, 12
Mass transfer factor, 11, 113
Mass transfer limited reaction, 15, 240
Mechanical strength
  of catalyst particles, 9, 23, 143
  effect of preparation on, 105
  importance of, 33, 131
  loss in steam reforming, 192
  measurement of, 143
Mechanisms, in catalyst design, 88
Mesopores, 9, 11
  measurement of, 152
Metals
  alkali, 50
  alloys, 193
  catalytic activity of, 26, 50, 62
  cluster compounds, 63
  gauzes, 123
  poisoning of, 205
  porous, 121
  porphorins, 220
  rare earth, 50
  transition, 50
Metathesis, 240
Methanation, 240
  catalyst stability in, 25
  loss of nickel, 193
  reactors for, 25
  role of nickel, 25
  temperature profiles in, 203
Methane, oxidation of, 85
Methanol, aromatic formation from, 81
Methanol synthesis
  by metals, 50
  steam reforming in, 25
Micropores, 9
Mid-boiling point, 240
Middle distillate, 240
Modeling
  of deactivation, 188
  of pilot units, 47
Monocycloparaffins, 240
Monolayer
  determination of, 147

Monolayer (cont.)
  formation on metals, 163
  of hydrogen in Ni/SiO_2
Monolithe, 123
Monomer, 240
MoO_3
  dipersion by P, 39
  in hydrodesulfurization catalysts, 38
  loss during regeneration, 193
Mordenite, structure of, 77
Motor octane number (MON), 240

Naphtha, 240
Naphthenes, 241
Natural gas liquids (NGL), 241
Neutron diffraction, 138
Ni/Al_2O_3
  catalyst structure of, 32, 121
  concentration effects, 30, 31
  for naphtha reforming, 20
  $Ni[Al_2O_4]$ formation in, 117
  recipe for, 361
  surface areas of, 32
$Ni[Al_2O_4]$
  formation of, 111, 117
  preparation of, 107
Nickel
  ethylene adsorption on, 52, 58
  hydrogen adsorption on, 58
  loss during methanation, 193
  nucleation of, 118
  porphorin contamination, 219
  reactions catalyzed by, 62
  in steam reforming, 24
NiO/Al_2O_3, preparation of, 106
Ni/SiO_2
  crystallite size distributions in, 196
  hydrogen chemisorption on, 164
Nitrogen, poisoning of acids, 208
N_2O decomposition
  activity patterns for, 68
  by semiconductors, 65
Nucleation, in precipitation, 97

Olefin alkylation, 7
Orbitals, surface geometry of, 57
Osmium, reactions catalyzed by, 62
Overlayers, 186
Oxidation, 241, 255
  by metals, 27, 50

Oxidation (*cont.*)
 of methane, 85
 by oxides and sulfides, 27
 partial, 28
Oxides
 as catalysts, 70
 cemented, 122
 fused, 122
 mixed, 122
 preparation of dual, 106
 as supports, 31
Oxychlorination, 241
Oxydehydrogenation, 241
Oxygenolysis, 241, 255

Packed beds
 pressure drop in, 8
 uniform flow in, 8
Packing density, 142
Palladium, reactions catalyzed by, 62
Paracrystallinity, 121
Partial oxidation, 241
 of CH_4, 85
Particle
 density, 141
 failure, 191
 growth in precipitation, 97
 size distribution, 97, 191
Passivation, of reduced metal catalysts, 120
Pattern of behavior
 ammonia synthesis, 17
 CO oxidation, 90
 cumene dealkylation, 75
 dehydration, 69
 dehydrogenation, 69, 90
 ethanol decomposition, 69
 nitrous oxide decomposition, 68
 oxygen mobility, 89, 90
 reduction, 90
Pellet density, 11
Pellets, manufacture of, 127
Peptization, 102
Personnel, training in pilot units, 47
Phosphorous
 -based zeolites, 82
 promotion of hydrodesulfurization catalysts, 39
Photocatalysis, by semiconductors, 68
Physical adsorption, 146

Pilot units
 in catalyst manufacturing, 124
 testing with, 46
Platinum
 anion adsorption, 113
 for automobile exhaust catalysis, 28
 for catalytic reforming, 28, 34
 crystallite size effects, 30, 199
Poisoning
 agents, 201
 beneficial, 209
 with diffusion, 204
 of metals, 205
 origin of, 200
 of oxides, 207
 titration, 202
 treatment of, 207
 uniform, 201
Poison titration, 162
Polymer, 241
Polymerization, 252
 by insulators, 27
Polynuclear aromatics, 241
Pores
 ink bottle, 218
 macro, 10
 meso, 9
 micro, 9
 origin of, 10
Pore shape
 determination of, 154
 importance of, 9
Pore size distribution
 bimodal, 39
 importance of, 9
 measurement of, 151
 nomenclature for, 9
 origin of, 10
Pore structure, control of, 218
Pore volume
 loss on drying, 103
 nomenclature for, 11
Porosimetry, 151
Porosity
 nomenclature for, 11
 particle, 13
 in supports, 31
Potassium
 in catalytic cracking, 214
 in hydrodesulfurization, 39

Potassium (cont.)
    in methanation, 37
    as a promoter, 36
    in steam reforming, 48, 93
Pour point, 241
Powders, manufacture of, 131
Power rate law, 18, 173
Precipitation
    of the active component, 109
    effect of pH, 98
    of supports, 97
Preparation, of supports, 95
Pressure drop
    effect of particles, 10
    in packed beds, 8
Process
    design, 48
    development, 42
    diagnosis, 19
    need for a catalyst, 41
    zeolites in, 82
Process variables, tested in pilot units, 46
Promoters
    function of, 35
    to prevent deactivation, 190
    in processes, 36
    site blocking, 60
Pt/Al₂O₃, 31
    crystallite migration in, 196
Pulse reactor, types of, 179

Radial crushing strength, 143
Raffinate, 241
Raney metals, 121
Rate constants, for activity, 173
Rate equation
    activation energy for, 21
    for activity, 172
    apparent order, 21
    first order, 174
    global, 18
    Langmuir-Hinshelwood, 174
    power law, 18
Rates
    for activity, 172
    effect of mixing speed, 179
    effect of particle size, 20
    effect of velocity, 22
    regimes, 20
Reaction
    redox, 26

Reaction (cont.)
    steps, 12
    structure sensitive, 53
    surface, 16
    thermodynamic feasibility of, 3
    titration, 162
Reactors
    differential, 175, 177
    experimental, 174
    gradientless, 178
    integral, 175
    loading, 132
    pulse, 179
    recirculation, 179
    temperature profiles in, 190
    tubular, 175
Redox reactions, 26
Reduction
    of the catalyst, 117
    passivation of Ni after, 120
    pre-, 120
Reductive alkylation, 241
Refining, 241
Reformate, 241
Regeneration
    of catalysts, 188
    effect of heavy metals, 192
Research octane number (RON), 241
Residua
    bimodal pores for hydrodesulfurization, 217
    hydrodesulfurization of, 39
    poisoning effect of heavy metals, 192
Reynolds number, 11, 175
Rhenium
    promotion by, 216
    reactions catalyzed by, 62
Rhodium, reactions catalyzed by, 62
Rings, pressure drop with, 10
Ruthenium, reactions catalyzed by, 63

Scale-up
    of catalyst recipes, 124
    data from pilot units, 47
Scanning electron microscopy (SEM), 158
Selectivity, 241
    benefits of, 5
    effect of catalyst, 4
    effects of crystallite size, 199
    shape and size, 75, 78, 80
Semiconductors
    catalysis by, 26

Semiconductors (*cont.*)
  doping of, 39
  electronic band structure of, 63, 67
  intrinsic, 63
  n-type, 65
  oxides, 28
  p-type, 65
  sulfides, 28
  types of, 66
Sensitivity, 241
Shape selectivity
  in coking, 212
  in zeolites, 75, 78, 80
Shift reaction, 241
Side chain, 241
Silica
  gels, 33, 100
  recipe for, 259
  as spheres, 129
  as a structural promoter, 193
  surface area of, 148
Silica alumina
  acidity of, 168
  acid sites in, 73
  carbonium ion reactions with, 74
  coking on, 211
  manufactured as spheres, 129
  metal porphorin contamination in, 218
  particle size distribution, 143
  potassium promotion of, 214
  preparation of, 106
  recipe for, 260
Silver, reactions catalyzed by, 62
Sintering
  of colloidal metals, 29
  in deactivation, 194
  mechanisms of, 195
  of supports, 195
Skeletal density, 141
Small angle x-ray scattering, 160
Smoke point, 241
Sodium, as a promoter, 36
Solid acids, reactions catalyzed by, 27, 75
Sols
  adsorption of, 109
  gelation of, 101
  preparation of, 97
Sour or sweet crude, 241
Specific rate, 172
Spinel
  Al$_2$O$_3$ as a defect, 105

Spinel (*cont.*)
  reactions between, 108
Spheres
  manufacture of, 129
  pressure drop with, 10
Stability, of silica gels, 100
Steam cracking, 242
Steam reforming, 242
  catalyst requirements for, 30
  coking in, 221
  collapse of pores in, 195
  deactivation in, 189
  loss of mechanical strength in, 192
  of methane, 24
  of naphtha, 20, 33
  loss of potassium in, 48, 193
  process demands on, 43
  loss of silica in, 43
Straight run gasoline, 242
Substitute natural gas (SNG), 21, 28, 242
Sulfides
  as catalysts, 70
  as semiconductors, 66
Sulfiding, of hydrodesulfurization catalysts,
    120
Supersaturation, in sol precipitation, 97
Supports
  function of, 28
  metal interactions with, 60
  oxides as, 31
Surface
  accessibility, 162
  dilution in alloys, 59
  mechanisms in catalyst design, 89
  metal sites, 61
  morphology, 158
  structure of, 161
Surface area
  apparatus for measuring, 149, 150, 151
  external, 12
  importance of, 9
  measurement of, 9
  nomenclature, 11
Surface diffusion, 16
Synthesis gas, 242

Tablets, pressure drop with, 10
Tamm temperature, 29
Temperature programmed reduction (TPR),
    140

Texture
    features of, 9
    parameters, 10
Theoretical density, 141
Thermodynamics
    in catalyst design, 87
    effect of catalysts, 1
    feasibility from, 3
Thiele modulus, 13, 203
TiO$_2$, for CH$_4$ oxidation, 91
Toluene disproportionation, 80
Tortuosity, 13, 155, 157
Toxicity
    of metals, 206
    of nonmetals, 206
Transalkylation, 242
Transmission electron microscopy (TEM), 158
Turnover number, 173

Ultraviolet photon spectroscopy (UPS), 161
Ultraviolet spectroscopy, 157
Unit operations, in catalyst manufacturing, 125
Urea, for precipitation, 110

Vanadium, porphorin contamination, 219
Visbreaking, 242
V$_2$O$_5$, for CH$_4$ oxidation
Volcano curves, 16, 17, 53, 55
Vol'kenstein theory, 68
Voorhies equation, 212

Washcoat, 123
Washing
    after deposition, 111
    of hydrogels, 102
Wicke–Kallenback method, 155
Weight hourly space velocity (WHSV), 242

Xerogels
    pellet formation from, 104
    preparation of, 102
X-ray diffraction (XRD)
    particle size determination, 159
    phase identification, 139
X-ray fluorescence, 137
X-ray photon spectroscopy, 161
Xylene isomerization, 80

Yield, 242

Zeolites
    coking on, 211
    origin of, 74
    preparation of, 108
    structure of, 76
Zeta potential, in preparation, 100
ZnO, for CH$_4$ oxidation, 91
ZnTiO$_3$, for CH$_4$ oxidation, 91
ZSM-5
    shape selectivity in, 80
    structure of, 77